8th International Conference on Environmental Science and Technology (ICEST 2017)

IOP Conference Series: Earth and Environmental Science
Volume 78

Madrid, Spain
12 – 14 June 2017

ISBN: 978-1-5108-4716-3
ISSN: 1755-1307

Printed from e-media with permission by:

Curran Associates, Inc.
57 Morehouse Lane
Red Hook, NY 12571

Some format issues inherent in the e-media version may also appear in this print version.

Printed by Curran Associates, Inc. (2017)

For permission requests, please contact the Institute of Physics
at the address below.

Institute of Physics
Dirac House, Temple Back
Bristol BS1 6BE UK

Phone: 44 1 17 929 7481
Fax: 44 1 17 920 0979

techtracking@iop.org

Additional copies of this publication are available from:

Curran Associates, Inc.
57 Morehouse Lane
Red Hook, NY 12571 USA
Phone: 845-758-0400
Fax: 845-758-2633
Email: curran@proceedings.com
Web: www.proceedings.com

Table of contents

Volume 78

8th International Conference on Environmental Science and Technology (ICEST 2017)

12–14 June 2017, Madrid, Spain

Accepted papers received: 20 July 2017
Published online: 21 August 2017

Preface

Environmental Pollution Control and Resource Management

Environmental Biology and Environmental Biotechnology

Energy Engineering and Management

IOP Conf. Series: Earth and Environmental Science **78** (2017) 011001 doi:10.1088/1755-1315/78/1/011001

PREFACE

It is our great pleasure to welcome you to 2017 2017 8th International Conference on Environmental Science and Technology (ICEST 2017) which will be held in Computer Science School (Escuela Técnica Superior de Ingenieros Informaticos, ETSIInf) of Technical University Of Madrid (UPM), Madrid, Spain during June 12-14, 2017. ICEST 2017 is dedicated to issues related to Environmental Science and Technology.

The major goal and feature of the conference is to bring academic scientists, engineers, industry researchers together to exchange and share their experiences and research results, and discuss the practical challenges encountered and the solutions adopted. Professors from Spain, Australia, United Kingdom, USA are invited to deliver keynote speeches and plenary speeches regarding latest information in their respective expertise areas. It will be a golden opportunity for the students, researchers and engineers to interact with the experts and specialists to get their advice or consultation on technical matters, sales and marketing strategies.

This proceedings present a selection from papers submitted to the conference from universities, research institutes and industries. All of the papers were subjected to peer-review by conference committee members and international reviewers. The papers selected depended on their quality and their relevancy to the conference. The volume tends to present to the readers the recent advances in the field of Environment and Industrial Innovation and various related areas, such as Advanced Ceramic-porous ceramic, Aerodynamics, Application of Spatial technology, Catalysis and Environment, Enivronmental Biotechnology, Environment management, Environmental Chemistry, Environmental engineering, Environmental Planning of Mines, Environmental Pollution Control, Fermentation Technology, Industrial Environment, etc..

We would like to thank all the authors who have contributed to this volume and also to the organizing committee, reviewers, speakers, chairpersons, sponsors and all the conference participants for their support to ICEST 2017.

Prof. Roberto San Jose
Technical University of Madrid (UPM), Madrid, Spain
July 4, 2017

Conference Committee

Conference General Co-Chairs

Prof. Roberto San Jose, Technical University of Madrid (UPM), Madrid, Spain Prof. Carlos Garbisu, The Basque Institute for Agricultural Research and Development, SpainProf. R. J. (Dick) Haynes, The University of Queensland, St Lucia, Queensland

Program Co-Chairs

Prof. Graciela Metternicht, University of New South Wales AustraliaProf. Bogdan Zygmunt, Faculty of Chemistry, Gdansk University of Technology, Poland

Publication Chair

Prof. Carlos Garbisu, The Basque Institute for Agricultural Research and Development, Spain

Technical Committee

Prof. Muscolo Adele, Mediterranea University, Italy
Prof. Gordon Huang, University of Regina, Canada
Prof. Vladimir Strezov, Macquarie University, Australia
Prof. Marek Trojanowicz, Institute of Nuclear Chemistry and Technology, Poland
Prof. Hasan Arman, United Arab Emirates University, UAE
Prof. Vasilis Fthenakis, Center for Life Cycle Analyis Columbia University, USA
Prof. Reza Nekovei, Texas A&M University-Kingsville, USA
Prof. Selda Tekin Özan, Süleyman Demirel University, Turkey
Prof. Aysegul Tanik, Istanbul Technical University, Turkey
Prof. Woo-Kyun Lee, Department of Environmental Science and Ecological Engineering, Korea University, South Korea
Prof. Zakaria HAMIMI, Benha University, Egypt
Prof. LIN Chuxia, University of Salford, Greater Manchester, UK
Assoc. Prof. Jinsheng You, School of Natural Resources, University of Nebraska, Lincoln, USA
Dr. Tae-Hyoung Tommy Gim, Seoul National University, Korea
Dr. Jen-Jeng Chen, Tajen University, Taiwan
Dr. Sherif Ali Younis Sarhan, Egyptian Petroleum Research Institute, Egypt
Dr. Roohollah Noori, University of Tehran, Iran

IOP Conf. Series: Earth and Environmental Science **78** (2017) 011003 doi:10.1088/1755-1315/78/1/011003

Peer review statement

All papers published in this volume of *IOP Conference Series: Earth and Environmental Science* have been peer reviewed through processes administered by the proceedings Editors. Reviews were conducted by expert referees to the professional and scientific standards expected of a proceedings journal published by IOP Publishing.

Acoustic Agglomeration Process of Fine Particles in a Resonance Structure

Chen-hao SHI [1], Jian ZHANG [2], Yun ZHAO [3], Jie LIANG [1]

1 College of Environmental Science and Engineering, Hunan University, Changsha, 410082, China
2 School of Information Science and Engineering, Central South University, Changsha, 410083, China
3 School of Optoelectronic Science and Engineering, National Defense Science and Technology University， Changsha, 410073, China

Abstract: It was proved that the acoustic agglomeration technology has a good application prospect in the removal of fine particles. In this paper, a removal system of acoustic agglomeration is constructed by the acoustic resonance structure. With the finite element simulation model, the effect and condition of sound pressure level (SPL) increment of high intensity sound in the resonance structure are defined. In the experiment, the contrast of the sampling weight and particle size distribution changes of fine particles was compared under different operating conditions to examine the effect of acoustic agglomeration on the removal efficiency of fine particles. The results show the SPL increment of 10dB is obtained with SPL 145-165 dB when the working frequency is changed from 400 to 2000 Hz. Under the action of acoustic agglomeration, fine particles in the aerosol were significantly reduced, and the removal effect is markedly improved with the increase of SPL.

1. Introduction

In recent years, the fine particles, mainly PM2.5, have become the main pollutant of urban atmospheric in China 0. In the current air pollution control, it is urgently necessary to develop industrial scale fine particles emission reduction technology for the combustion energy system. The removal method based on the acoustic agglomeration principle has a good application prospect, due to its short working time, significant effect, easy to use, and high temperature, high pressure and corrosive prevention. However, the complex mechanism of acoustic agglomeration, lack of high-power and high intensity sound source and high energy consumption are the three major bottlenecks in the practical application of the acoustic agglomeration technology 0.

There have been many theoretical and experimental works on acoustic agglomeration0, but most of their experimental platform is bulit on imitative conventional dust removal devices combined with an acoustic agglomeration system. Furthermore, the results are not totally consistent because of the complex mechanisms and different experimental conditions. Zhou etc. 0used sound wave to work on conventional dust removal devices and found that the optimized sound frequency and SPL were 1.4 kHz and 142dB. Gallego0compared the effect of 10 kHz and 20 kHz sound wave on flue gas generated by a coal-fired fluidized bed and proved that 20 kHz sound source was better than 10 kHz for agglomeration.Wang Jie0compared high-frequency (1 kHz and 20 kHz) acoustic wave with low-frequency (0.5~3 kHz) acoustic wave and found that low-frequency acoustic wave did better than the high one working on coal fired flue gas. Although there was no common view on the optimized conditions because of the varied investigated operating conditions, overall the works proved the effect of sound wave.

However, few works have been done to indicate the positive effect of sound wave with the professional dust removal equipment based on strong sound system, which is exactly closely related to the actual industrial application.Conventional acoustic agglomeration tests are usually carried out in plane traveling wave or standing wave condition. The existing experimental results show that the acoustic agglomeration efficiency of the polydisperse aerosols for the industrial flue gas normally has the optimal agglomeration frequency, and the agglomeration efficiency is significantly enhanced with the increase of SPL. Therefore, replacing the constant section traveling wave/standing wave tube by the acoustic resonance cavity0, and as the acoustic resonance frequency is equal to the optimal frequency, the removal of fine particles under the condition of acoustic structure resonance can reduce the sound power consumption at the premise of keeping high reduction efficiency.

2. Experimental Platform

Figure 1 shows the test system of acoustic agglomeration of fine particles based on acoustic resonance structure. The system consists of a standing-wave tube, which is composed of thin tube and square tube in series, where the cross-section dimension of square thick tube is 0.3 m with its length 0.5 m, and the cross-section length of thin one (condensed tank) is 0.1 m with its initial length 1.5 m. The thin tube is connected with the piston (changing the variable position) to adjust the length of the tube, and then change the resonance frequency of the standing wave tub. The standing wave tube is driven by four electro-acoustic unit, which is connected with the thick tube. Electro-acoustic unit is composed of 150 W, 400-5000 Hz compression driver and conical horn. Inlet and outlet diameters of the horn are 0.05 m and 0.1 m respectively, and its length is 0.2 m (Dip angle is approximately 7 deg.). The frequency and intensity of sound wave are regulated by a signal generator and power amplifier. The amplitude of driving current of a single sound source is 3 A at full capacity.

Compared with commonly used constant section standing wave tube, the two-stage standing-wave tube with abrupt varying section can be used to improve the effect of acoustic agglomeration in the thin tube from two aspects: (1) one end of the thick tube is allowed to connect more compression drivers to provide higher input power. (2) The standing-wave tube with abrupt varying section is a kind of dissonant standing-wave tube, i.e., the high order resonance frequency is unequal to an integer multiple of the first order resonance frequency. Therefore, under the condition of the structural resonance and the high intensity acoustic field, the nonlinear distortion of the acoustic wave is lower, and sound saturation and shock wave will not occur generally.

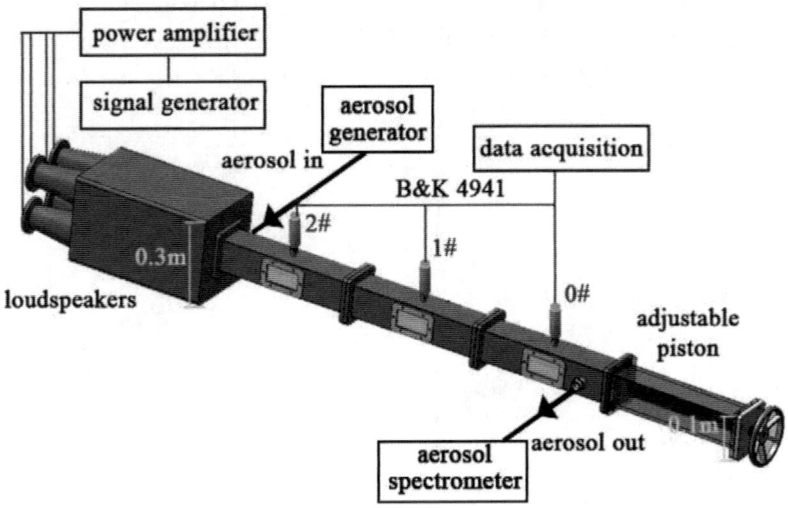

Figure 1. .Acoustic agglomeration system.

According to the waveguide theory, the cut-off frequency of acoustic agglomeration chamber with length of 0.1m is 1715 Hz, which can cover the effective frequency band (400-1500Hz) for acoustic agglomeration of fine particles. According to the law of quarter wavelength, the adjustable piston stroke is 0.5 m, which is more than half wavelength and in the adjustment range of the resonance frequency in the working frequency band.

The input gas to the agglomeration chamber is generated by the SAG-410 Topas aerosol generator, and the range of mass concentration is 0.012-13 g/m^3. When the gas is mixed with the compressed air, the range of gas volume flow is 7.2-18 m^3/h, and the range of corresponding action time by high intensity acoustic wave is 3-7 s. Three B&K 4941 high intensity acoustic microphones, 0#, 1# and 2#, are installed uniformly along the thin tube, which locate at point of the 0.5 m long tube, respectively. 0# is near the end of the adjustable piston, and the 2# is near the end of the sound source. The acoustic microphones are connected with the B&K 3050 data acquisition system to obtain the acoustic signals of measuring point. A particle size spectrometer is installed at the gas outlet of agglomeration chamber, which is used to measure the particle size distribution characteristic of the particles with and without the sound waves, and to complete the sampling of fine particles. The flue gas outlet is connected with a fine particulate sampling instrument, and the particle size of the sample is collected ranging from 2.5 to 10 μm by sampling membrane filter.

3. Strong Sound Field Characteristics

3.1. *The Time Domain Signal and Frequency Response*

Using the frequency domain acoustic module of Comsol Multiphysics, the acoustic field distribution in the experimental system is calculated. The results show that three high order resonance frequency of the system is 1570 Hz, 1510 Hz and 1420 Hz respectively. The SPL of agglomeration chamber is lowest when the frequency is 1550 Hz. When the structure resonance occurs, the SPL in the thin tube is above 10 db comparing with that in the thick tube. The internal acoustic field distribution at the frequency of 1420 Hz is shown in Figure 2.(a), it can be seen that there is a plane standing wave acoustic field in the agglomeration chamber, and the acoustic field distribution is very inhomogeneous in the thick tube, and sound intensity of thin tube (see Figure 2. (a)) is obviously higher than that of thick tube (see Figure 2. (b). A similar standing wave acoustic field is observed at the other two resonance frequencies.

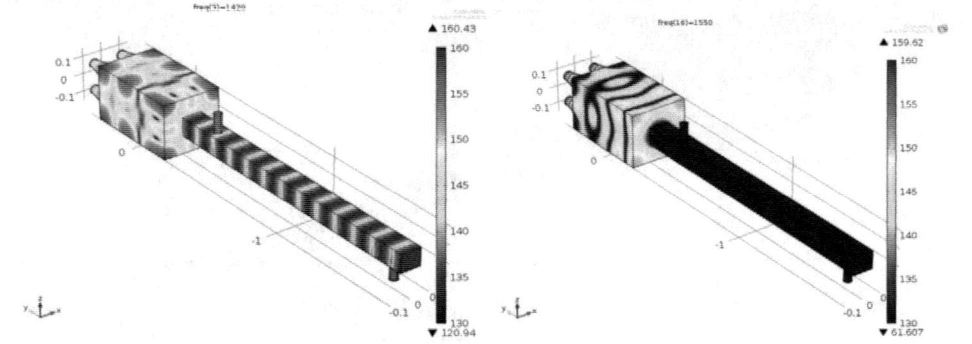

(a) 1420Hz (b) 1550Hz

Figure 2. Acoustic field distribution under the resonance condition.

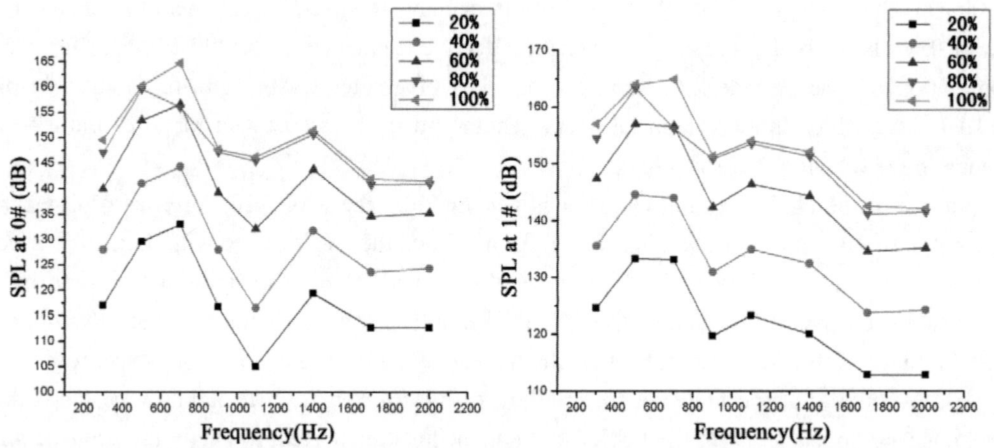

(a) 0# (b) 1#

Figure 3. The RMS SPL of every measuring points at different frequency and sound intensity.

Figure 3 shows the RMS sound pressure level of 0# and 1# monitoring point close to the adjustable piston at different frequency and intensity of the sound source. It can be seen that under the effect of standing wave field, the SPL fluctuations within the working frequency band of compression driver increased significantly. The SPL is highest when the frequency is 700 Hz, and is increased by more than 10 dB compared with the results of single source driver, which indicates that phase congruency of different sound source is better. When the drive current of compression driver is changed from 20% to 80%, the frequency response curve shape and the increase amplitude of SPL basically exhibits no change. When the agglomeration effect is efficiently in the range of 400 - 1500 Hz, the loudest level is as high as 150-165 dB. Compared with the results of Figure 2, the peak value of sound pressure frequency response at fixed monitoring point could not reach the maximum sound pressure level of the agglomeration cabin. At the same time, because the structural resonance frequency is not chosen as the measurement frequency, the measurement results of Figure 3 are lower than the actual intensity of the acoustic agglomeration process.

3.2. Resonance Frequency and Harmonic

Resonance frequency of the standing-wave tube with abrupt varying section is determined by the length and sectional area of thick tube and thin one, adjusting individually the length of agglomeration chamber can make the resonance frequency be consistent with the sensitive frequency of acoustic agglomeration. When the position of adjustable piston is changed, the resonance frequency of standing wave tube will change, and the relative position of the measuring point in the sound field also changes.

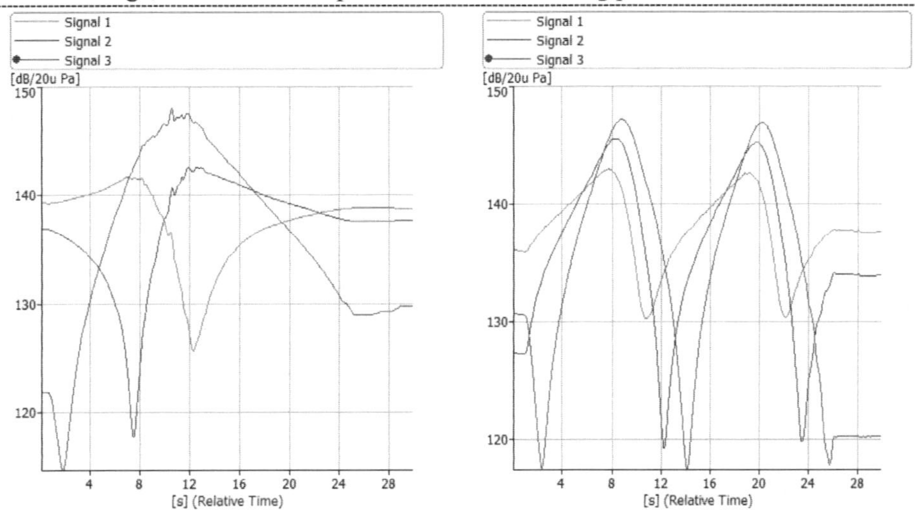

(a) 600Hz (b) 1400Hz

Figure 4. The RMS SPL of every measuring point when the adjustable piston position is changed and the driving current is 40%.

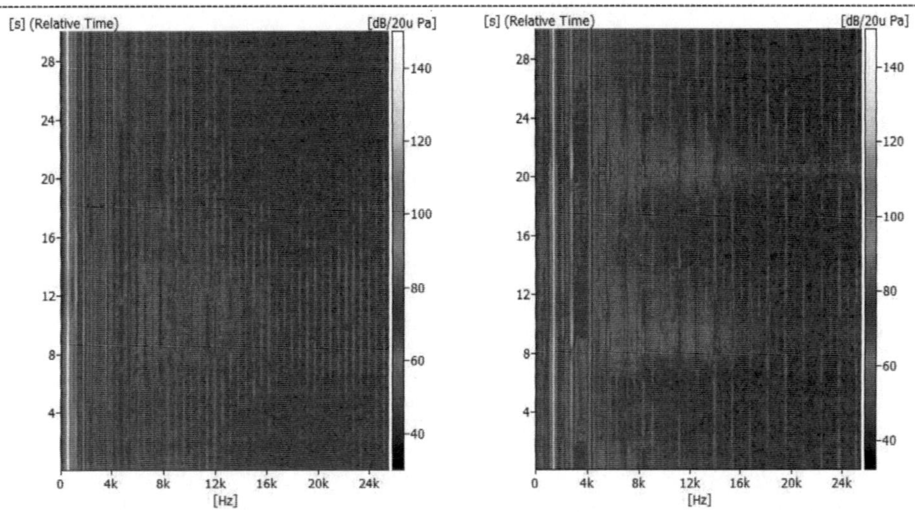

(a) 600Hz	(b) 1400Hz

Figure 5. The RMS SPL of measuring point 2 when the adjustable piston position is changed and the driving current is 40%.

Figure 4 shows the change of SPL of measuring point when the piston position is changed and the amplitude of driving current at full capacity is 1.2 A. When the frequency is 600 Hz, SPL of the measuring point in the curve reaches a maximum value, which is higher than the SPL measured at the initial position of piston. The data of measuring point 0# and 2# are basically the same, and the time moment of the maximum SPL of 0# and 2# is different with that of 1#. When the frequency is 1400 Hz, because the piston stroke is greater than the acoustic wavelength, the SPL periodically changes over time. The time moment of the maximum SPL at different points is different. Similarly, the maximum SPL is higher than that of the initial position of the piston.

Figure 5 shows the change of spectrum over time at 1#when piston position is changed and the amplitude of driving current at full capacity is 1.2 A. It can be seen that the main energy appears at the fundamental frequency, at the same time, there is an obvious high order harmonic, where harmonic component becomes more significant when the SPL is higher.

The maximum SPL of each test point driven by single sound source (fixed piston) or 4 sound sources (fixed and adjustable piston) are shown in Figure 6 It can be seen that the SPL driven by 4 sound sources is 10 dB more than that by the single source, and the phase congruency of different sound sources is better. Moving the piston to a proper position, the effect of resonance will occur from 4 acoustic sources, which improve the loudest SPL level. The intensity of the acoustic signal recorded under the fixed piston and fixed measured point condition is actually less than the intensity of the acoustic wave in the process of the acoustic agglomeration.

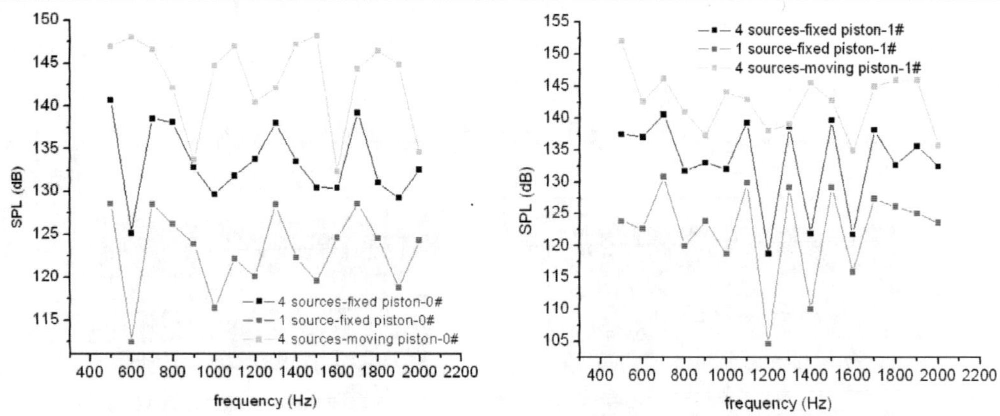

(a) 0# (b) 1#

Figure 6. The RMS SPL at fixed and mobile piston measuring point when the driving current is 40%.

4. Fine Particle Coagulation Results

The effect of acoustic agglomeration varies with the frequency and intensity of sound waves. Figure 7 shows the change of mass of PM2.5 in the flue gas under different frequency sound waves. The PM2.5 concentration in the initial distribution of flue gas is 49.64 mg/m^3, and the agglomeration effect of fine particles in the high intensity acoustic field is obvious. When the frequency is lower, the content of fine particles in the flue gas is higher, while that in the range of 1000 - 1300 Hz is the smallest. However, when the frequency is further increased, the content is increased. That is, for flue gas generated by coal-fired fly ash, the optimal frequency for sound agglomeration is in the range of 1000 - 1300 Hz.

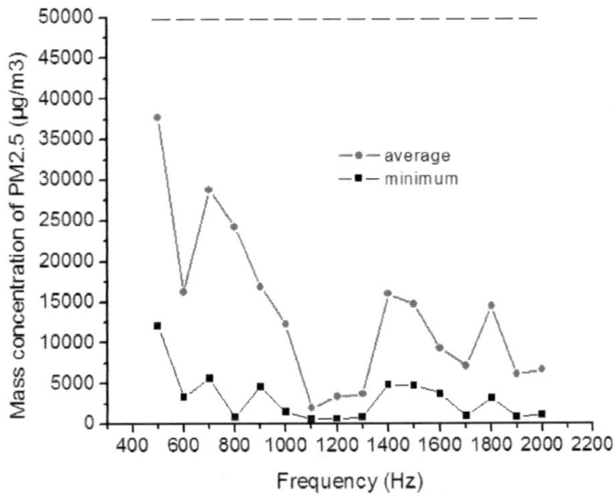

Figure 7. Variation of PM2.5 mass concentration in flue gas under different frequency sound waves.

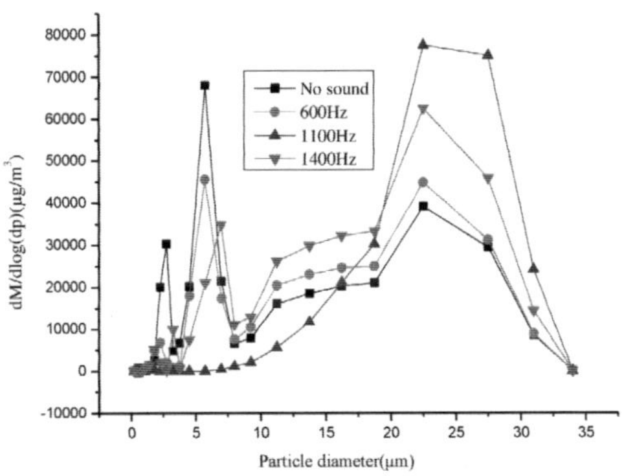

Figure8. Variation of particle size distribution (mass) of particles in flue gas by 600 Hz, 1100 Hz and 1400 Hz sound waves.

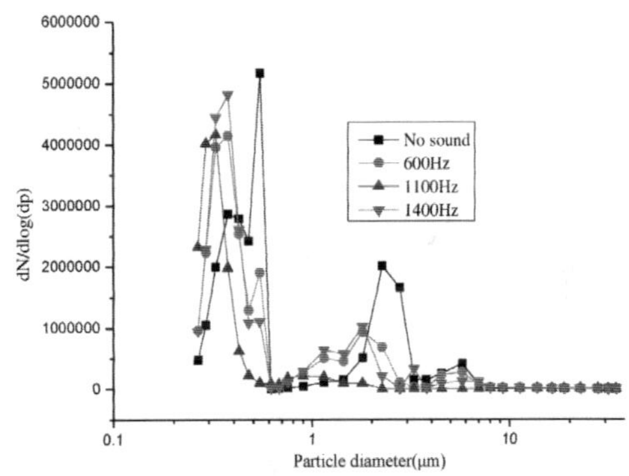

Figure9. Variation of particle size distribution (quantity) of particles in flue gas by 600 Hz, 1100 Hz and 1400 Hz sound waves.

Figure 8 and Figure 9 give the variation of the mass and the number of different size particles in the flue gas under the action of 600 Hz, 1100 Hz and 1400 Hz, respectively. It can be seen that there are a large number of fine particles in the initial distribution of flue gas, and the particle number distribution showed a typical bimodal structure. There are more numbers of particles with about 6 μm in size, and the most numbers of particles with about 0.5 μm and 2 μm in size. In the results of the initial gas particle mass distribution, the weight of the particles about 10 μm and 2 μm in size are larger, and the particle weight of 6 μm in size is the largest. At the optimal frequency of 1100 Hz sound waves, the mass of particles below 5 μm in size decreased significantly, and the mass of particles above 10 μm in size has a certain increase, which proves the effect of particles agglomeration is active. In the results of number distribution, the number of particles in the range of 0.5 to 2 μm in size decreases sharply, however, the number of particle 0.3 μm in size has a certain increase. In addition, under the actions of the sound waves at the frequency of 600 Hz and 1400 Hz, which are far away from the optimal frequency, the agglomeration effect of particles is significantly decreased.

5. Conclusion

In order to provide high efficiency and low energy consumption of removal system of fine particles based on the acoustic agglomeration technology, the acoustic resonance structure is designed based on the standing-wave tube with abrupt varying section, which replaces the conventional experimental system. The experimental tests and numerical computations are carried out to study the acoustic properties and the agglomeration process of fine particles. The following results were obtained:

- In the range of the sensitive frequency (1000 Hz - 2000 Hz) of acoustic agglomeration of the fine particles, there are multiple resonance frequencies in the standing-wave tube with abrupt varying section, and the SPL increment of 10 dB is obtained when structure resonance occurs. In the range of frequency from 500 Hz to 2000 Hz, the SPL in the agglomeration chamber can reach 145 to 165dB.
- The particle size distribution in the flue gas generated by the fly ash from the power plant is a typical bimodal structure. Under the action of high intensity sound waves within the measured frequency band, the mass of the particles below 5 μm in size is decreased significantly, and the mass of the particles above 10 μm in size has a certain increase. Especially at the sensitive frequency of 1100 - 1300 Hz, the mass aggregation of PM2.5 is decreased by more than 90%.
- The acoustic energy in the agglomeration chamber appears at the fundamental frequency, and the harmonic components are reduced by the resonance of the acoustic wave in the standing wave tube. Shock wave caused by the nonlinear effect is not found in the experimental results.

In the future, we shall connect our acoustic agglomeration system with the bag filters in the power plant to test the overall effect of removal of PM2.5 particles.

6. References

[1] Hoffmann T L. Environmental implications of acoustic aerosol agglomeration [J]. Ultrasonics,2000, 38(1–8):353-357.

[2] Du X, Kong Q, Ge W H, Zhang S J, Fu L X. Characterization of personal exposure concentration of fine particles for adults and children exposed to high ambient concentrations in Beijing, China. Journal of Environmental Sciences, 2010, 22(11): 1757–1764.

[3] Lin-jun Yang.Combustion source of fine particulate matter pollution control technology[M].Chemical industry press, 2011.32-67.(in Chinese)

[4] Zhou D, Luo Z, Jiang J, et al. Experimental study on improving the efficiency of dust removers by using acoustic agglomeration as pretreatment[J]. Powder Technology, 2016: 52-59.

[5] Yao gang, Zhao bing, xiang-lin Shen. Coal particulate matter waves together effect of experimental research and numerical analysis. Thermal power engineering, 2006,21(2):175-178.

[6] Fan F, Chen H, Yuan Z, et al. Convergence of Fine Particles under Standing-Wave Condition[C]// International Conference on Bioinformatics and Biomedical Engineering. IEEE, 2010:1-4.

[7] G.X.Zhang, J.Z.Liu, J.B. Wang etc, Theoretical study of acoustic wake effect in acoustic

agglomeration, Journal of Chemical Engineering of Chinese Universities, 2013,27(2):199-204.

[8] J.A. Gallego-Juarez, et al., Application of acoustic agglomeration to reduce fine particle emissions from coal combustion plants, Environmental Science & Technology,1999,33(21): 3843-3849.

[9] Wang J, Liu J, Zhang G, Zhou J, Cen K. Orthogonal design process optimization and single factor analysis for bimodal acoustic agglomeration. Powder Technol. 2011; 210(3):315-22.

[10] Liu J Z, Zhang G X, Zhou J H,Wang J, Zhao W D, Cen K F. Experimental study of acoustic agglomeration of coal-fired fly ash particles at low frequencies. Powder Technology, 2009, 193(1): 20–25.

[11] Jianzhong Liu, Jie Wang, Guangxue Zhang, Junhu Zhou, Kefa Cen. Frequency comparative study of coal-fired fly ash acoustic agglomeration. Journal of Environmental Sciences, 2011,23(11):1845–1851.

[12] Min qi, Peng feng.Mutation section standing wave tube and high pure standing-wave field experimental research. Journal of acoustics, 2010,35 (2): 185-191.

[13] Lawrenson C C, Lipkens B, Lucas T S, et al. Measurements of macrosonic standing waves in oscillating cavities[J]. Journal of the Acoustical Society of America, 1997, 102(5):623-636.

[14] Ilinskii Y A, Lipkens B, Lucas T S, et al. Nonlinear standing waves in an acoustical resonator[J]. Journal of the Acoustical Society of America, 1998, 104(5):2664-2674.

[15] Ilinskii Y A, Lipkens B, Zabolotskaya E A. Energy losses in an acoustical resonator.[J]. Journal of the Acoustical Society of America, 2001, 109(5 Pt 1):1859.

Modelling the Effect of Black Carbon and Sulfate Aerosol on the Regional Meteorology Factors

X Ma[1], W Wen[2]

1 National Meteorological Center, CMA, Beijing 100081, RP China
2 Institute of Urban Meteorology, China Meteorological Administration, Beijing 100089, China
E-mail addresses: brianwenwei@163.com

Abstract. In this study, we focus on the effect of black carbon aerosol and sulfate aerosol on meteorology factors during heavy pollution period and non-heavy pollution period. The version of WRF/chem V3.4 was used in this work, Four Simulation scenarios are applied to simulate the effect of the effect of black carbon aerosol and sulfate aerosol on solar radiation, temperature, PBL high. The analysis results show that the effect of black carbon and sulfate aerosol cause decline on three meteorological factors in both heavy pollution and non-heavy pollution period in both January and July. The influence of two aerosols on meteorological factors are less significant than winter. During heavy pollution, black carbon aerosol cause the loss of solar radiation is 29.1W/m2; the warming effect of black carbon aerosol caused temperature to rise 0.05℃; PBL height decreased by an average of 73.1m. Sulfate aerosols cause the loss of solar radiation is 21.5W/m2; Temperature fell an average of 0.89℃; PBL height decreased by 66.6m .The change of three meteorological factors due to aerosol feedback in non-heavy pollution period in much smaller than heavy pollution period.

1. Introduction

China has become one of the most uncertainty areas in aerosol radiative forcing and climate change due to the severe air pollution. The effect of aerosol radiative force is different among different aerosol species. Sulfate aerosol refers to the form of solid particles suspended in the air of sulfuric acid or sulfuric acid, with cooling effect. Black carbon (BC) aerosol is the main component of soot, by solid fuels (coal, wood, dung and crop residues), biomass combustion and combustion of fossil fuels into the atmosphere, both cooling and heating effect. Global anthropogenic contributions to the concentration of aerosols (primarily sulfate, organic carbon, black carbon, nitrate and dust) together produce a cooling effect with a total (including atmosphere and surface) direct radiative forcing of -0.5 W/m2 and an indirect cloud albedo forcing of -0.7 W/m2[1,2,3,4]. Simulation of the aerosol-radiation interaction and aerosol-cloud interaction of aerosol feedbacks require on-line coupled models [5].

The aerosol-radiation interaction and aerosol-cloud interaction of aerosols on the regional meteorological factor in the short-term are important in the dispersion of pollutants and in secondary organic reactions. A five day period study with Weather Research and Forecasting model with Chemistry (WRF/Chem) with the Model of Aerosol Dynamics, Reaction, Ionization, and Dissolution (MADRID) (referred to as WRF/chem-MADRID) of 2000 Texas Air Quality (TexAQS2000) found that the incoming solar radiation and near-surface temperature decreased due to aerosol-radiation interaction and aerosol-cloud interaction feedbacks, and aerosols led to excessive numbers of cloud condensation nuclei with too low saturation and reduced precipitation via these CCN [6,7].

Among the previously described studies, field experiments have deficiencies in investigating the aerosol radiation interaction and aerosol-cloud interaction of aerosols on regional meteorological factors and air quality over a large region and a long-term (years and decades) period. Although using general circulation models and climate models can focus on the regional and long-term period aerosol feedbacks, but it cannot fully reflect the short-term (hours and days) characteristic of aerosol feedbacks. On-line air quality models are designed for aerosol simulation, which include the accurate simulation of the aerosol concentration and the regional- to global-scale of the feedback between air pollutants and meteorological factors[5]. In this study, we focus on the effect of black carbon aerosol and sulfate aerosol on meteorology factors during heavy pollution period (While $PM_{2.5}$ concentration higher than 250μg/m^3 or AQI higher than 150) and non-heavy pollution period (While $PM_{2.5}$ concentration lower than 250μg/m^3 or AQI lower than 150). Six typical cities (Beijing, Tianjin, Shijiazhuang, Shanghai, Nanjing and Guangzhou) are selected to analysis the effect in winter and summer in 2010.

2. Methodology

Four Simulation scenarios are applied in this study, as shown in table 1. The BASE scenario represent the real atmosphere, both aerosol-radiation interaction and aerosol-cloud interaction processes are included. The Effect_Non scenario does not calculate any aerosol feedback process. The BC_Non scenario does not include the effect of BC. The $SO4^{2-}$_Non scenario does not include the effect of sulfate aerosol. The effect of BC and sulfate aerosol are calculated based on the follow method.

Aerosol Effect = BASE - Effect_Non

BC Effect = (BASE - Effect_Non) - (BC_Non - Effect_Non)

Sulfate Effect = (BASE - Effect_Non) - ($SO4^{2-}$_Non - Effect_Non)

Table 1. Simulation scenarios set up

Scenarios	Description
BASE	Real atmosphere scenario, all aerosol feedbacks are included
Effect_Non	No aerosol feedback
BC_Non	No black carbon aerosol`s effect
$SO4^{2-}$_Non	No sulfate aerosol`s effect

3. Domain Setting and Model Configuration

The version of WRF/chem V3.4 was aplied in this work. The study area is divided into 92 × 78 horizontal grid cells with a 54-km grid spacing. The entire China continent, part of East China Sea and part of Southeast Asia are included. 27 logarithmic structure layers divide the modelling vertical zone, which range from the surface to a fixed pressure of 100mb. The National Centers for Environmental Prediction (NECP) Final Analysis Reanalysis data are used to generate the meteorology initial conditions and boundary conditions. The INTEX-B emission inventory for Asia in 2006 was used as the anthropogenic emission input[8]. Natural emissions are calculated online based on the U.S. Geological Survey (USGS) land use data. The meteorological observed data come from the Meteorological Information Comprehensive Analysis and Process System (MICAPS).

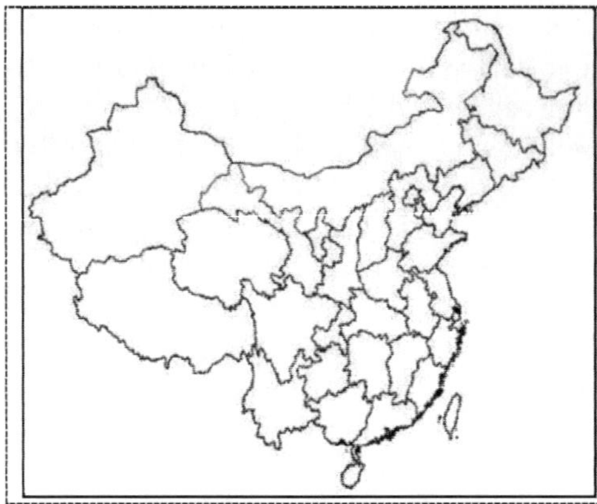

Figure 1. Simulation domain

The summarised performance statistics of the meteorology factors and the PM_{10} concentration levels are listed in table 2. figure1 compares the simulated and observed hourly T2 at Beijing, Shanghai, Wuhan and Guangzhou, which are located in China in the north, south, central and east, respectively, WRF/chem well captured the variations of T2 at all sites in the four months with correlation coefficients (RC) between 0.78~0.93. Warm bias occurs in some periods, which represent -0.50℃ to 2.16℃ MB and -1.71% to 44.08% NMB of four sites in the four seasons, especially in April. The warm bias is caused by the limitations in the PBL scheme, radiation schemes[9] and the land-surface model[10].The moderate to large over-prediction occurred in the hourly WSP10 in all the sites, with a domain wide average bias are 0.77, 1.35, 0.41, 0.69 m/s in January, April, July and October, respectively. The bias is caused by two points. First, analysis nudging (FDDA) was not applied to the meteorological factors to prevent the aerosol-radiation interaction and aerosol-cloud interaction of aerosols on meteorological factors are the suppressed [11]. Second, with a horizontal resolution of 54 km, the model has difficulties in capturing the wind properly. The hourly precipitation is well captured by WRF/chem, but incorrect reports (which represent cases when rain did not occur, but the model reported it and vice versa) of precipitations are found in the north China sites, and under-predictions are found in the south China sites.

$$NMB = \frac{\sum_1^N (c_m - c_0)}{\sum_1^N c_0} \times 100\% \qquad (1)$$

$$NMB = \frac{\sum_1^N |c_m - c_0|}{\sum_1^N c_0} \times 100\% \qquad (2)$$

Table 2. Simulation error analysis of PM_{10}

	City	NMB(%)	NME(%)		City	NMB(%)	NME(%)
	Beijing	-68.07	74.44		**Beijing**	23.28	50.71
	Shijiazhuang	-51.99	52.79		**Shijiazhuang**	3.62	28.91
	Tianjin	-50.52	50.52		**Tianjin**	14.89	32.91
January	**Shanghai**	-0.22	35.41	**July**	**Shanghai**	-39.35	42.07
	Zhengzhou	-23.21	30.54		**Zhengzhou**	26.88	45.19
	Wuhan	-21.68	31.02		**Wuhan**	-39.77	51.86
	Guangzhou	-41.95	52.44		**Guangzhou**	-56.24	57.8
	Average	-36.8	46.7		**Average**	-9.5	44.2

4. Results and Discussion

The result of the effect of BC and sulfate aerosol on meteorological factors is show in figure 2 and figure 3.

4.1. *Solar Radiation*

Solar radiation is decreased in six cities due to the effect of aerosol feedback in both winter and summer in 2010 in six cities, as shown in the first row in figure 2 and figure 3. The absorption of BC is stronger than sulfate`s scattering on solar radiation in both winter and summer. In January 2010, black carbon aerosol decline solar radiation and the average drop of 21.9W/m^2, heavy pollution period average drop is 29.1W/m^2, non-heavy pollution period drop is 14.6W/m^2. Sulfate aerosols cause the loss of solar radiation and the average drop is 14.8W/m^2, heavy pollution period is 21.5W/m^2, non-heavy pollution period drop is 10.6W/m^2. In July 2010, solar radiation decrease 12.1W/m^2 due to the effect of aerosol. The declining contents are similar during heavy pollution period and non-heavy pollution period. Black carbon aerosol and sulfate aerosols on solar radiation effects are relatively mild, average respectively
0.15 W/m2 and 0.9 W/m2

4.2. *Temperature*

Temperature is declined due to the effect of aerosol in six cities. In January 2010, the temperature fell an average of 0.89℃ in heavy pollution period and fell an average of 0.2℃ in non-heavy pollution period, as shown in the second row in figure 2 and Figure. The warming effect of black carbon aerosol caused temperature to rise 0.05℃. During the heavy pollution period, the temperature increased by 0.6℃, and during non-heavy pollution period temperature increased by 0.34℃, as shown in the right most column in figure 2. The cooling effect of sulfate aerosol resulted in an average reduction of 0.17 ℃ in six cities, and the temperature during the heavy pollution period was reduced by an average of about 0.43℃ and 0.08℃ during the period of non-heavy pollution. In July 2010, black carbon aerosol makes the temperature in the six cities increase slightly. Compared with January 2010, the effect of aerosol feedbacks on the temperature in the six cities are relatively light, and the effect of black carbon aerosol and sulfate aerosol on temperature are not significant.

4.3. *Plant Boundary Layer (PBL) High*

In January 2010, the effect of aerosol feedback make the average PBL high in the six cities decreased by 25.9m, and the PBL high decreased by 54.6m in the heavy pollution period; 17.2m in non-heavy pollution period, as shown in the third row in figure 2 and figure 3. Black carbon aerosol made six areas of PBL height decreased by an average of 36.4m, decreased 73.1m during the heavy pollution period and decreased 0.27m during the non-heavy pollution period. Sulfate aerosol made the average height of PBL in six cities decreased by 32.2m, decreased by 66.6m during the heavy pollution period, and increased by 2.3m during the non-heavy pollution period. In July 2010, the aerosol feedbacks makes the height of PBL in the six regions decreased by an average of 20.6m. In both heavy pollution period and non-heavy pollution period, and the variation of PBL height is not significant. Black carbon aerosol made PBL average height increased by 1.4m, sulfate aerosol had little effect on the PBL high. Compared with January 2010, the effect of aerosol on the PBL high in six cities are relatively light, the influence of black carbon aerosol and sulfate aerosol on PBL height are not significant too.

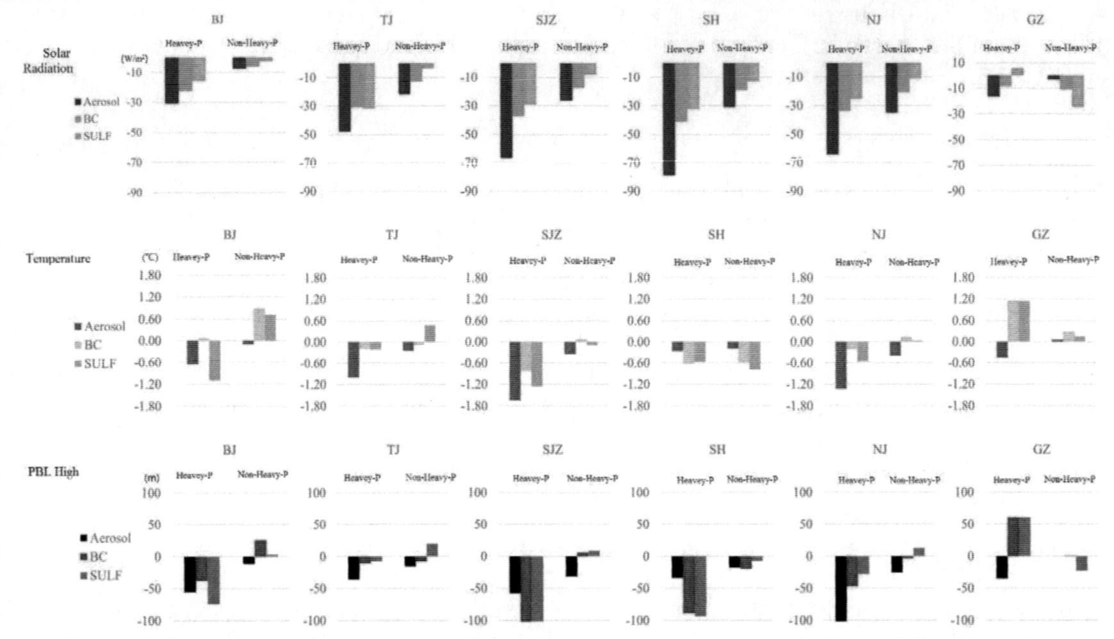

Figure 2. The effect of all aerosol, black carbon and sulfate on solar radiation, temperature and PBL high in January 2010

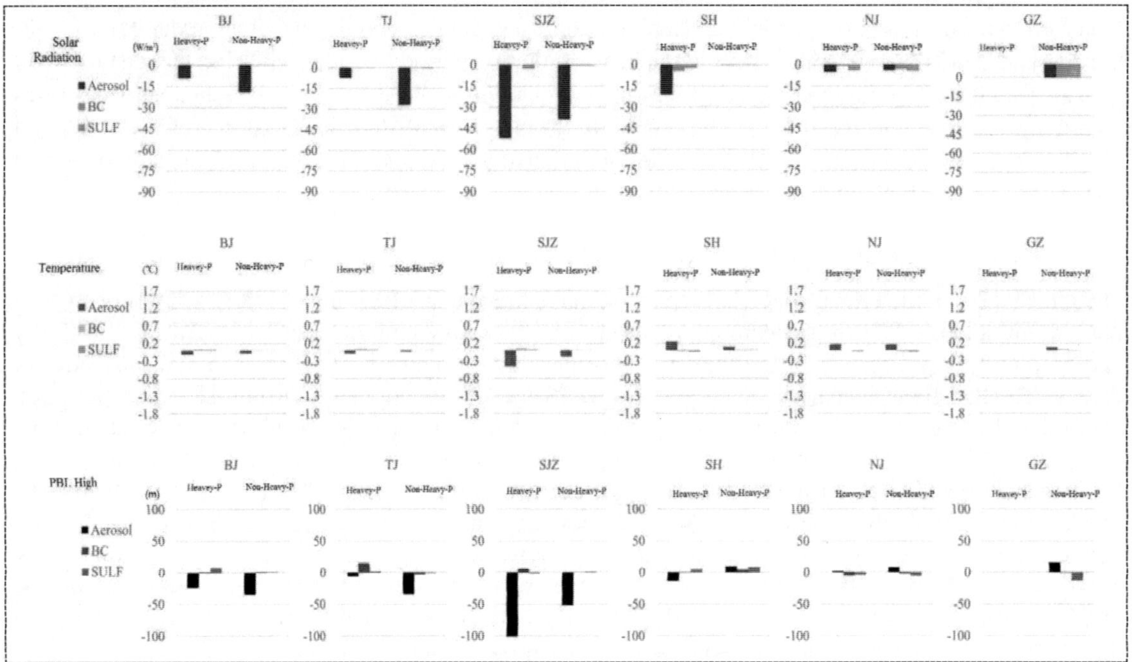

Figure 3. The effect of all aerosol, black carbon and sulfate on solar radiation, temperature and PBL high in July 2010

5. Conclusions

In this paper, the effects of black carbon aerosol and sulfate aerosol on solar radiation, temperature and PBL height in six cities over China in winter and summer are analyzed. According to the classification standard of air pollution index (AQI), When the $PM_{2.5}$ concentration is higher than 250 g/m^3, it is regarded as the heavy pollution period, lower than 250 g/m^3 is considered as non-heavy pollution period. The analysis results show that the absorption effect of black carbon aerosol is stronger than the scattering effect of sulfate on solar radiation. The warming effect of black carbon aerosol is significant in both heavy pollution and non-heavy pollution period. Both black carbon and sulfate aerosol are caused PBL height decreased. In this study, we only discuss the effect of black carbon and sulfate aerosol on solar radiation, temperature and PBL high. For further studies, the effect of the change in regional meteorological factors on the formation of secondary aerosols due to aerosol feedbacks.

6. Acknowledgments

This research was supported by the youth fund of the development of GRAPES model special fund (GRAPES-FZZX-2016-31) and Beijing Natural Science Foundation（8172051）.

7. References

[1] IPCC, Climate change 2007: Synthesis Report. In: Solomon, S., Qin, D., Manning, M. *Contribution of Working Group I to the Fourth Assessment Report of the Intergovernmental Panel on Climate Change. 2007b*

[2] Ramanathan V., P. J. Crutzen, J. T. Kiehl, D. Rosenfeld. Aerosols, Climate, and the Hydrological Cycle. *Sciences*, 2010. 5549: 2119-2124.

[3] Myhre, G., D. Shindell, F.-M. Bréon, W. Collins, J. Fuglestvedt, J. Huang, D. Koch, J.-F. Lamarque, D. Lee, B. Mendoza, T. Nakajima, A. Robock, G. Stephens, T. Takemura and H. Zhang, Anthropogenic and Natural Radiative Forcing. In: Climate Change 2013: The Physical Science Basis[R]. *Contribution of Working Group I to the Fifth Assessment Report of the Intergovernmental Panel on Climate Change [Stocker, T.F., D. Qin, G.-K. Plattner, M. Tignor, S.K. Allen, J. Boschung, A. Nauels, Y. Xia, V. Bex and P.M. Midgley (eds.)].* Cambridge University Press, Cambridge, United Kingdom and New York, NY, USA. 2013

[4] Jacobson, M.Z., (2002). *Atmospheric Pollution: History, Sciences and Regulation*, Cambridge University Press, New York, ISBN 0521010446, 399 pp.

[5] Zhang Y. Online-coupled meteorology and chemistry models: history, current status, and outlook, *Atmos. Chem. Phys.*, 2008, 8, 2895–2932.

[6] Zhang Y, Ying Pan. Kai Wang. Jerome D. Fast. Georg A. Grellb. WRF/Chem-MADRID: Incorporation of an aerosol module into WRF/Chem and its initial application to the TexAQS2000 episode. *Journal of Geophysical Research-Atmospheres.* 2010, 115, D18202

[7] Zhuang BL, Jiang F,Wang TJ, Li S，Zhu B. Investigation on the direct radiative effect of fossil fuel black-carbon aerosol over China. *THEORETICAL AND APPLIED CLIMATOLOGY.* 2011, 104, 301–312

[8] Q.Zhang, D.G.Streets, G.R.Carmichael, K.B.He, H.Huo, A.Kannari, Z.Klimont, I.S.Park, S.Reddy,J.S.Fu, D.Chen, L.Duan, Y.Lei, L.T.Wang, Z.L.Yao. Asian emissions in 2006 for the NASA INTEX-B mission. *Atmos. Chem. Phys.*, 2009, 9, 5131–5153.

[9] M. Garc´ıa-D´ıez, Fernandez. J, Fita. L, Yaguee.C. Seasonal dependence of WRF model biases and sensitivity to PBL schemes over Europe. *Quarterly Journal of the Royal Meteorological Society*. 2012, 139: 501–514

[10] Lee, S.-H., Kim, S.-W., Angevine, W. M., Bianco, L., McKeen, S. A., Senff, C. J., Trainer, M., Tucker, S. C., and Zamora, R. J. Evaluation of urban surface parameterizations in the WRF model using measurements during the Texas Air Quality Study 2006 field campaign. *Atmos. Chem. Phys.*,2011, 11, 2127-2143

[11] Renate Forkel, Johannes Werhahn, Ayoe Buus Hansen, Stuart McKeen, Steven Peckham, Georg Grell, Peter Suppan. Effect of aerosol-radiation feedback on regional air quality - A case study with WRF/Chem. *Atmospheric Environment*. 2012, 53, 202-211.

The Impact of Meteorological Factors on PM2.5 Variations in Hong Kong

X Li[1,2], Y J Feng[1], H Y Liang[2]

1 Shandong Agricultural University, Tai'an, Shandong 271000, China
2 The Hong Kong Polytechnic University, Hong Kong 999077, China
vivianyoxi@gmail.com

Abstract. This paper aims to explore the impact of meteorological factors on PM2.5 concentrations in Hong Kong. The PM2.5 and meteorology data including temperature, pressure, rainfall, relative humidity (RH), wind speed and wind direction from January to December of 2013 were collected. The correlation analysis between PM2.5 and meteorological factors were conducted for each month and season. The meteorology data were classified into several intervals and the mean PM2.5 concentrations for each interval were calculated to see the tendency. According to the correlation analysis results, the PM2.5 concentrations have a positive relationship with pressure with correlation coefficients 0.507 while have negative relationships with temperature, RH, rainfall, wind speed with correlation coefficients -0.512, -0.237, -0.524, -0.284, respectively. In addition, the wind direction influence PM2.5 concentrations through affecting the spreading direction of PM2.5. The north wind in winter increased the PM2.5 in Hong Kong while the south wind in summer decrease the PM2.5. Therefore, the meteorological factors affect the aggregation, diffusion, spread of PM2.5. They have a leading impact on PM2.5 concentrations when the domestic emission stays stable.

1. Introduction

The rapid development of urban construction and accelerated urbanization have caused serious pressure on atmosphere environment. As one result of the process, air pollution has increased concerns about people's health, especially for the fine particular matter (PM2.5). PM2.5 is the particular matter with an aerodynamic diameter less than 2.5 μm, which not only attenuates atmospheric visibility, but also has an adverse impact on human health [1]. Some epidemiology studies reveal that a certain relationship exists between particular matter of pollution and the mortality rate, cardiovascular and respiratory diseases [2], [3].

The PM2.5 concentrations were mainly influenced by local pollution emission, external pollution propagation and meteorological conditions including pressure, temperature, humidity, cloud coverage, precipitation and wind, etc. The meteorological factors cause spatio-temporal variation of particulate matter concentrations through influencing the conglomeration and diffusion of pollutants. Tai [4] indicate that the daily variation of meteorological parameters can cause up to 50% of the change of PM2.5 concentrations. Wang and Ogawa [5] investigated the correlation between PM2.5 and meteorological conditions in Nagasaki, Japan. The results shown that temperature had a negative, and precipitation had a positive, correlation with PM2.5. Elminir [6] found that temperature, humidity, wind speed and wind direction contribute to the air pollutants concentration. Wind direction was found to have influence on both pollutant concentrations and the correlation between pollutants.

Hong Kong is located in southern China with sea on three sides and is one of the most densely populated areas. In the last decades, the economic cooperation between Hong Kong and Pearl River

IOP Conf. Series: Earth and Environmental Science **78** (2017) 012003 doi:10.1088/1755-1315/78/1/012003

Delta (PRD) area is growing rapidly. In addition, the increasing of population mobility and transportation brings enormous pressure to the local atmospheric condition.

Since the correlation between PM 2.5 concentration and meteorological factors varies with study area due to different regional climate conditions, the correlation model for other regions may not represent their relationship in Hong Kong. Moreover, due to the lack of heavy industry, the main sources of PM2.5 pollution in Hong Kong are vehicle exhaust, construction dust and other domestic sources [7]. Comparing with other areas suffering heavy pollution, such as Beijing, the impact of meteorological conditions on PM 2.5 variation in Hong Kong is more apparent. Therefore, it is essential to investigate this relationship in Hong Kong area.

This study utilized PM 2.5 concentration and meteorological data (including temperature, pressure, precipitation, humidity, wind speed and wind direction) during 2013 in Hong Kong, aiming to examine the impact of meteorological conditions on local PM2.5 variations. In this paper, the relationship between PM 2.5 pollution and local weather condition will be revealed, which will provide theoretical support for predicting PM 2.5 change based on meteorological parameters.

2. Data Collection

The daily averaged meteorological data including pressure (hPa), temperature (°C), relative humidity (%), rainfall (mm), wind speed (km/h), wind direction (°) from 1 Jan 2013 to 30 Dec 2013 at four automatic weather stations were collected from Hong Kong Observatory (HKO) (http://www.hko.gov.hk). Correspondingly, the hourly averaged PM2.5 concentrations data at four air quality monitoring stations which are most close with automatic weather stations in space were downloaded from Environmental Protection Department (EPD) website (http://www.epd.gov.hk/epd/). In order to be consistent with meteorological data in time unit, the daily averaged value of PM2.5 concentrations were calculated. Table 1 shows the locations of air quality stations and automatic weather stations.

Table 1. The location of PM2.5 monitoring stations and automatic weather stations

Air quality monitoring stations	Coordinate	Automatic Weather Stations	Coordinate
Sham Shuipo	114° 9' 33'' 22° 19' 49''	King's Park	114°10'22'' 22°18'43''
Shatin	114° 11' 4'' 22° 22' 35'	Shatin	114°12'36'' 22°24'09''
Tung Chung	113° 56' 37'' 22° 17' 20''	Sha Lo Wan	113°54'25'' 22°17'28''
Yun Long	114° 1' 22'' 22° 26' 43''	Wetland Park	114°00'32'' 22°28'00''

3. Result and Discussion

3.1. The Tendency of PM2.5 Concentration and Meteorological Factors

To explore the tendency of PM2.5 concentration and meteorological factors and their relationship, the monthly values of PM2.5 and meteorological factors are averaged from daily data. It's hard to compare all factors together due to their different units. For example, temperature's unit is °C and ranges from 10-30 °C while pressure's unit is hPa with values around 1000 hPa. Therefore, all kinds of data were processed by normalization and their values changes to 0-1 without units.

The monthly tendency of PM2.5 concentrations and normalized meteorological factors is shown in Figure1. According to Figure 1, the monthly average PM2.5 concentrations are lowest in summer (June, July, August), which are followed by spring (March, April, May) and autumn (September, October, November), and are highest in winter (January, February, December). Among all meteorological factors, only pressure has the similar monthly tendency with PM2.5 concentrations while temperature, rainfall, RH, wind speed and wind direction have opposite tendency with PM2.5.

In order to have a better knowledge of the correlation between PM2.5 and meteorological factors, the correlation coefficients of PM2.5-meteorology and each two meteorological factors were calculated from the daily average data in 2013. As shown in Table 2, PM2.5 concentrations have a positive correlation with pressure and negative correlation with temperature, rainfall, RH, wind speed and wind direction, which is consistent with Figure1. The correlation coefficients of PM2.5 concentration with temperature and relative humidity are -0.513 and -0.524 respectively, showing strong negative correlations. The correlation coefficients of PM2.5 concentration with rainfall and wind speed are -0.237 and -0.284 respectively. The PM2.5 concentration has a weak correlation with wind direction, which is -0.1 while has a strong positive correlation with the pressure of 0.507.

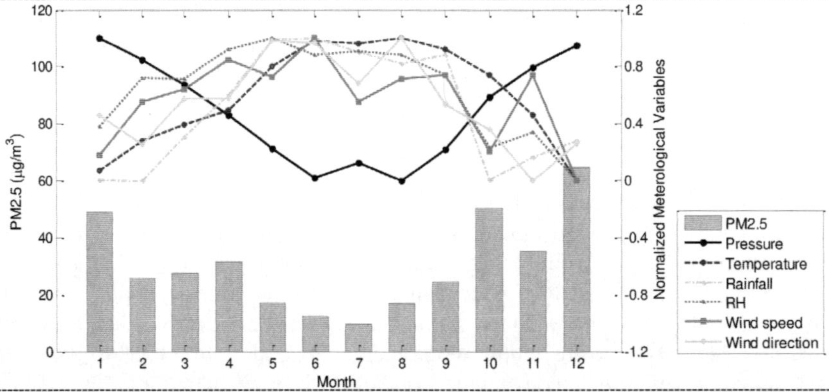

Figure 1. The monthly tendency of PM2.5 concentration and normalized meteorological factors

Table 2. The correlation coefficient between PM2.5 and meteorological factors in 2013

Correlation Coefficient	PM2.5	Temperature	Pressure	RH	Rainfall	Wind speed	Wind direction
PM2.5	1						
Temperature	-0.5132	1					
Pressure	0.5073	-0.8461	1				
RH	-0.5239	0.4002	-0.5244	1			
Rainfall	-0.2370	0.1219	-0.3133	0.3916	1		
Wind speed	-0.2843	0.1303	-0.1979	0.1128	0.0770	1	
Wind direction	-0.0997	0.2258	-0.2956	0.1460	0.1069	-0.0573	1

3.2. Correlations between PM2.5 and Temperature

According to the range of temperature in 2013, it is classified into 5 intervals which are 10-15, 15-20, 20-25, 25-30, 30-35. The amplitude of variation of temperature in Hong Kong is much smaller than that in high latitude area due to the subtropical monsoon climate. As shown in Figure 2, the low temperature 10-15 and high temperature 30-35 have low frequency, which are 8% and 2%, respectively. The highest frequency appears in the interval 25-30, which is 43%. The mean values of PM2.5 concentrations for each temperature intervals are calculated. The results show that PM2.5 concentrations decrease with temperature increase. The mean value of PM2.5 concentrations is highest (42 μg/m^3) when the temperature ranges between 10-15°C. When the temperature is higher than 25°C, PM2.5 mean concentrations become lower than 10 μg/m^3. This because when the temperature is higher, the air convection at lower surface is stronger, which benefits the upward transport of particular matter [8].

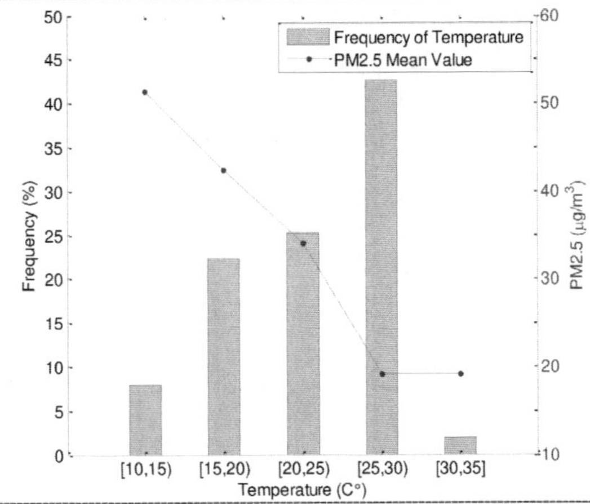

Figure 2. The frequency of temperature intervals and mean value of PM2.5 concentration for each interval

The linear regression between PM2.5 concentration and temperature was made for each season. As shown in Figure 3, the correlations of PM2.5 concentration and temperature are positive in winter, spring, and autumn. They have best linear regression fit in spring which R^2 is 0.23. The correlation is different in summer, which is positive with R^2 =0.022.

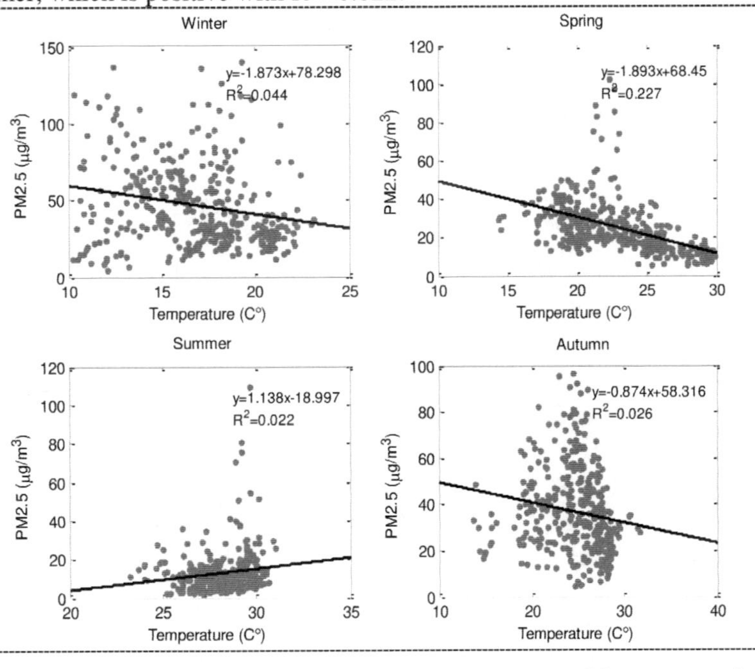

Figure 3. The linear regression of PM2.5 and temperature of four seasons in 2013

3.3. *Correlations between PM2.5 and Pressure*

The pressure in 2013 ranges from 990 hPa to1030 hPa. It was divided into 8 intervals: 990-995, 995-1000, 1000-1005, 1005-1010, 1010-1015, 1015-1020, 1020-1025, 1025-1030. According to Figure 4, pressure concentrates between 1005 hPa and 1020 hPa in most of the year. With pressure's increase, PM2.5 concentrations have a general increase tendency. Since when pressure is low, the surrounding air masses with high-pressure airflow to center. The upward airflow is formed in the center and helpful for the diffusion of pollutants. On the contrary, when the surface is dominated by high pressure, the center area has downward air flow. This condition is not helpful for the dilution and diffusion of pollutants and thus increase its concentration [9].

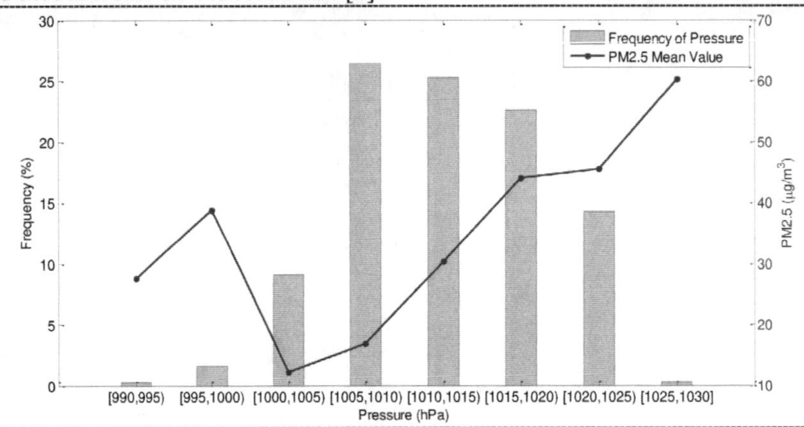

Figure 4. The frequency of pressure intervals and mean value of PM2.5 concentration for each interval

Figure 5 shows the linear regression analysis of PM2.5 concentration and pressure. In winter, spring and autumn, they have a positive correlation while in summer, they have a negative correlation.

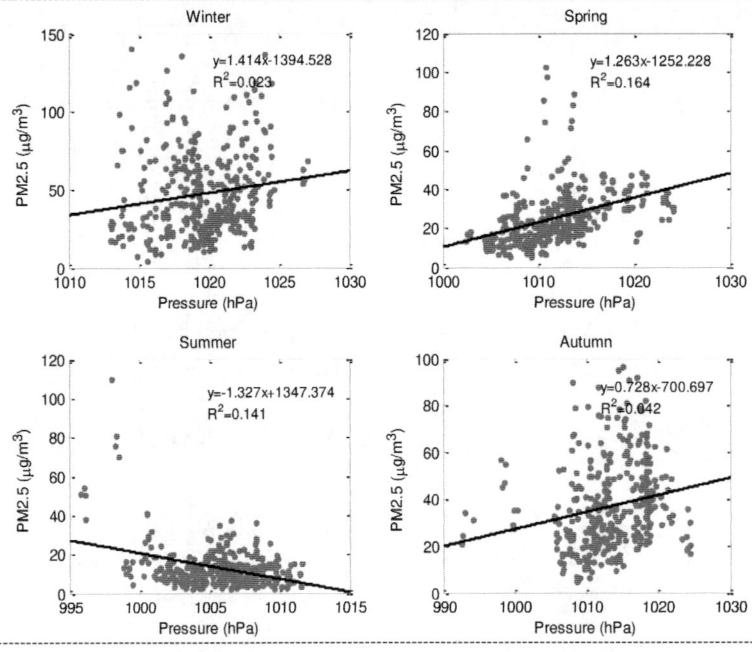

Figure 5. The linear regression of PM2.5 and pressure of four seasons in 2013

3.4. *Correlations between PM2.5 and RH*

The relative humidity in 2013 ranging from 20% - 100% was classified into 8 intervals: 20-30, 30-40, 40-50, 50-60, 60-70, 70-80, 80-90, 90-100. The frequency of each interval is shown by the histogram of Figure 6. Hong Kong has a humid climate and the frequency of RH between 70% and 90% exceeds 60%. The mean value of PM2.5 concentration in each RH interval is shown by the broken line in

Figure 6, which decrease with the increase of RH. Although there is an increase of PM2.5 when RH changes from 20-30 to 30-40, the frequency of RH in 20-30 is very low <1%. Therefore, the increase does not have universal representation.

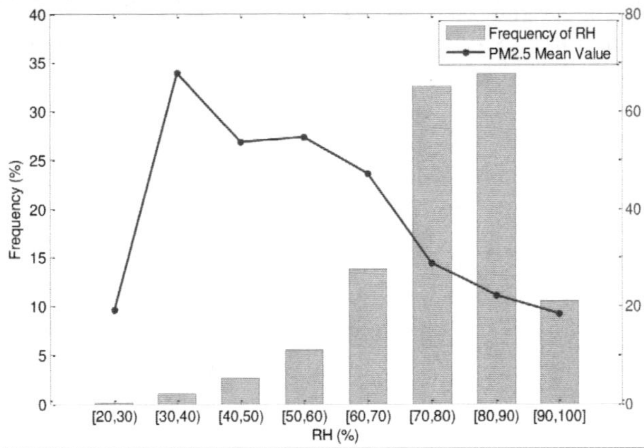

Figure 6. The frequency of RH intervals and mean value of PM2.5 concentration for each interval

Figure 7 shows the linear regression between PM2.5 concentration and RH. The PM2.5 concentrations and RH have negative correlations in winter, spring, autumn. They fit best in winter with R^2=0.247, followed by autumn R^2=0.114. In summer, PM2.5 doesn't have obvious correlations with RH (R^2=0.001).

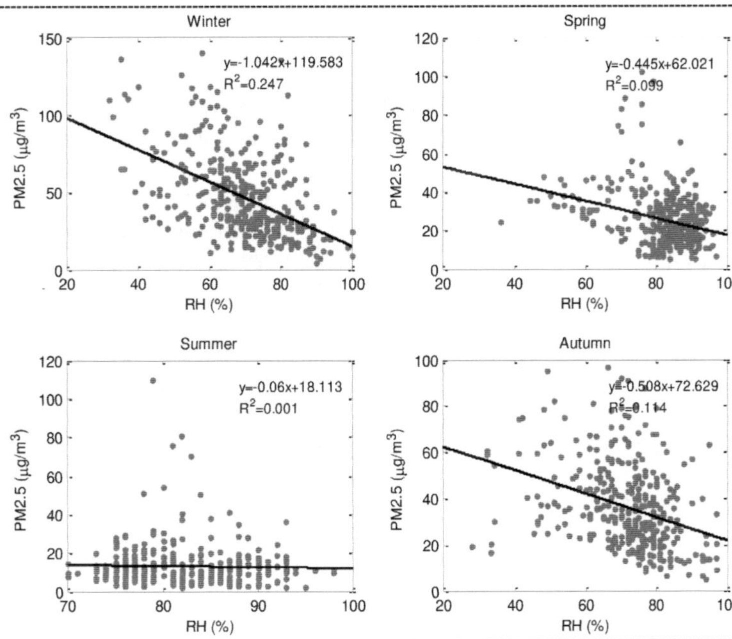

Figure 7. The linear regression of PM2.5 and RH of four seasons in 2013

3.5. *Correlations between PM2.5 and Rainfall*

Since rainfall mainly concentrates in summer, we don't make the correlations analysis of PM2.5 and rainfall for different seasons and months. We only analyze the PM2.5 tendency for different rainfall levels. The rainfall in 2013 were divided into12 intervals: <1, 1-10, 10-20, 20-30, 30-40, 40-50, 50-60,

60-70, 70-80, 80-90, 90-100, >100. The RH values for each interval were made statistics and the mean values of PM2.5 for each interval were calculated. As shown in Figure 8, the histogram shows the RH frequency and the broken line shows the PM2.5 variations tendency. About 70% days in 2013 have rainfall less than 1 mm. The frequency in 1-10 is 15%. The rainfall >70 mm has a very low frequency with 1%. Generally, the mean PM2.5 concentrations for different RH intervals decrease with rainfall. Especially when the rainfall changes from <1 to 1-10, PM2.5 has a big decline from 37 μg/m^3 to 22 μg/m^3. Therefore, the rainfall can clean the air and thus decrease the particular matter content.

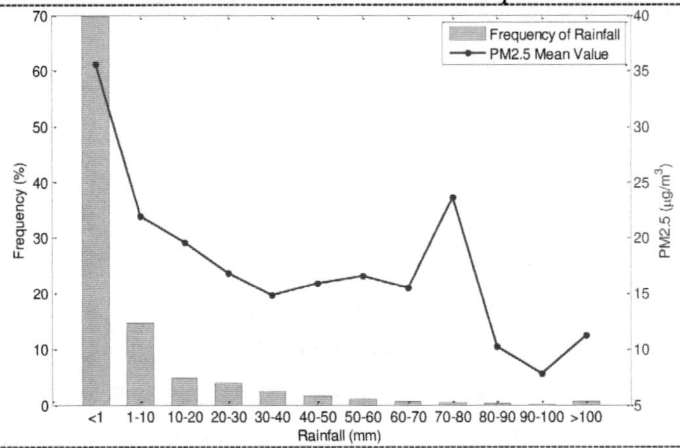

Figure 8. The frequency of rainfall intervals and mean value of PM2.5 concentration for each interval

3.6. *Correlations between PM2.5 and Wind Speed*

As an important meteorological factor, wind influence the horizontal and vertical transport of air pollutants. It also affects the speed of concentration and diffusion of pollutants directly [8]. The wind speed in 2013 is slow in general, ranging in 0-30 km/h. The wind speeds were classified into 6 intervals: 0-5, 5-10, 10-15, 15-20, 20-25, 25-30. The wind speeds of 5-10 km/h account for 60% of the whole year (Figure 9). The PM2.5 concentration reduces significantly with wind speed increases. When wind speed increase from 0-5 km/h to 5-10 km/h, the PM2.5 mean concentration has a largest reduce from 56 μg/m^3 to 34 μg/m^3. This is because high wind speed is beneficial for the diffusion of PM2.5 [10]. On the contrary, low wind speed inhibits the diffusion of PM2.5 and makes the PM2.5 concentrates in the surface.

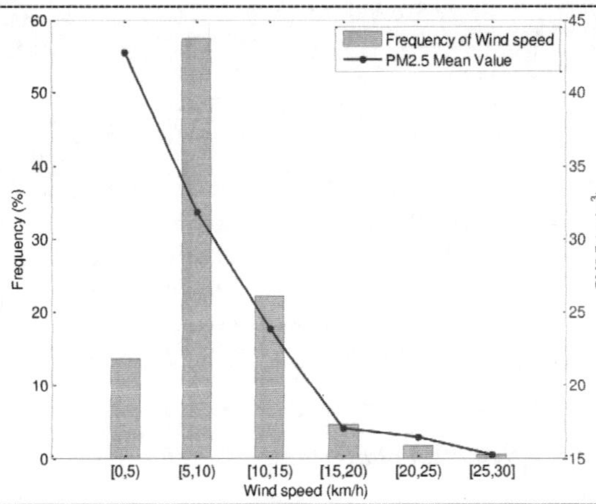

Figure 9. The frequency of wind speed intervals and mean value of PM2.5 concentration for each interval

According to Figure 10, the linear regressions between PM2.5 and wind speed are different in four seasons, but the slopes are all negative, which means PM2.5 decreases with wind speed increases.

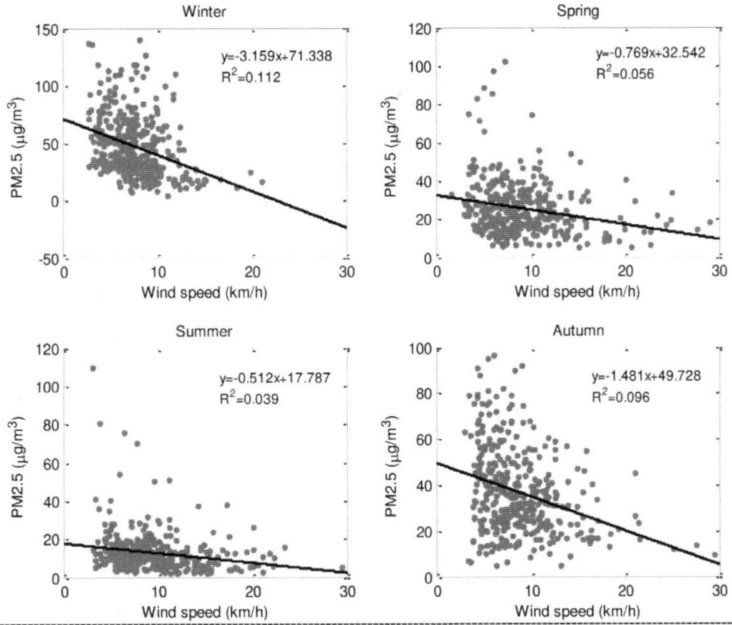

Figure 10. The linear regression of PM2.5 and wind speed of four seasons in 2013

3.7. *Correlations between PM2.5 and Wind Direction*

Besides wind speed, wind direction influences the transport direction of PM2.5 concentration [10]. The impact of wind speed on PM2.5 concentration was evaluated by Average Weighted PM2.5 (AWP), which represents the average PM2.5 concentration for each unit of wind speed. The AWP of one direction was calculated by the following equation [5]:

$$AWP = \frac{\sum_{i=1}^{N} P_i/WS_i}{N} \quad (1)$$

P_i is PM2.5 concentration, WS_i is wind speed, N represents the number of the direction.

Figure 11 depicts the AWP in four seasons of 2013. The AWP values for different directions are higher in winter ranging from 5μg/m^3 to 25 μg/m^3 than that in other three seasons. The AWP values (2-14 μg/m^3) are obviously lower in summer than that in other three seasons. Compared with the wind rose chart in Figure 12, winter is dominated by North wind with main directions of NNW, N, NE, ENE. The AWP of N wind is highest, larger than 20 μg/m3. The main reason is that the north wind in winter bring pollutants from Guangdong to Hong Kong and increase the PM2.5 concentration [11], [12]. The main winds in spring are southeast and northeast wind. The directions of NE, ENE, SSE, S have relative high frequency. The AWP values of NE and ENE wind (about 5 μg/m^3) are higher than that of SSE and S wind (about 4 μg/m^3). In autumn, east wind is the main direction. The frequency of directions ENE and E are 30% and 23%, respectively. The AWP values for all directions don't have a big difference. Summer was dominated by southeast wind with main directions of SSE, ENE, SSE (frequency is close to 40%). The AWP values for SSE and ENE are 2.1 μg/m^3 and 3.5 μg/m^3, respectively. This is because the southeast wind in summer brings clean air from ocean to Hong Kong [11].

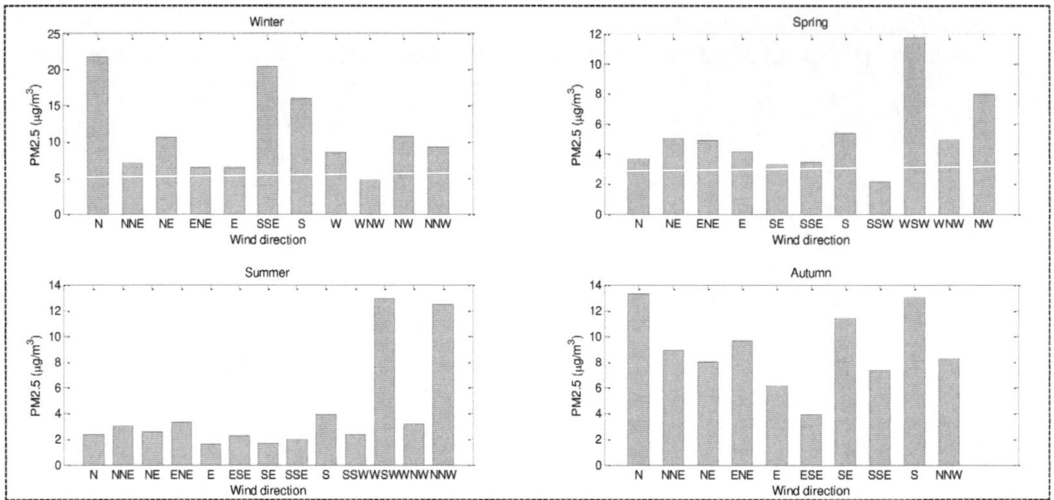

Figure 11. The AWP values of different wind directions in 4 seasons in 2013

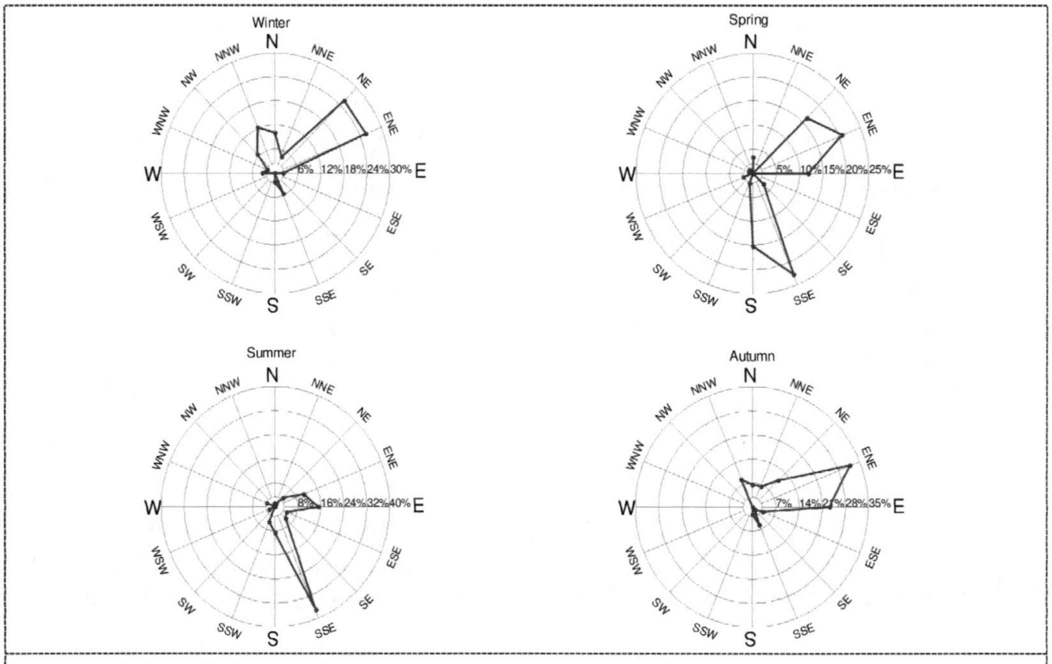

Figure 12. The wind rose of 4 seasons in 2013

4. Conclusion

This study has analyzed the systematic impact of meteorological factors on PM 2.5 variation based on PM 2.5 and weather data during 2013 in Hong Kong, the main findings of this study are concluded as follows:

- The PM2.5 concentration has positive correlations with pressure and negative correlations with temperature, rainfall, RH and wind speed. The correlation coefficients are highest with temperature, pressure, and RH, which indicates the three factors mainly influence the PM2.5 concentrations.
- The wind direction influences the transport direction of PM2.5 pollutants. The north wind in winter increases the PM2.5 concentration while the southeast wind in summer decreases the PM2.5 concentration.

- The correlations between PM2.5 and meteorological factors vary with different seasons. For example, PM2.5 has a positive correlation with temperature in winter and a negative correlation in summer.

In summary, meteorological factors have leading impacts on PM2.5 pollutions when the local emission is stable. In future studies, the correlation between PM2.5 concentration and meteorological factors can be used for prediction of air pollution and help government to make emission-control polices to prevent and reduce health impact of PM2.5.

5. Acknowledgments

Authors wishing to acknowledge Hong Kong Observatory and Environment Protection Department for the data providing.

6. Reference

[1] Shi W, Wong MS, Wang J, Zhao Y 2012 Analysis of airborne particulate matter (PM2.5) over Hong Kong using remote sensing and GIS. *Sensors* 12 6825.
[2] Qiu H, Yu ITS, Tian L, Wang X, Tse LA, Tam W, et al. 2012 Effects of coarse particulate matter on emergency hospital admissions for respiratory diseases: a time-series analysis in Hong Kong. *Environ. Health. Perspect.* 120 572.
[3] Peng RD, Bell ML, Geyh AS, McDermott A, Zeger SL, Samet JM, et al. 2009 Emergency Admissions for Cardiovascular and Respiratory Diseases and the Chemical Composition of Fine Particle Air Pollution. *Environ. Health. Perspect.* 117 957.
[4] Tai PKA. Impact of Climate Change on Fine Particulate Matter PM2.5 Air Quality 2012.
[5] Wang J, Ogawa S 2015 Effects of Meteorological Conditions on PM2.5 Concentrations in Nagasaki, Japan. *Int. J. Environ. Res. Public. Health.* 12 9089.
[6] Elminir HK 2005 Dependence of urban air pollutants on meteorology. *The Science of the total environment* 350 225.
[7] Ho KF, Cao JJ, Lee SC, Chan CK 2006 Source apportionment of PM2.5 in urban area of Hong Kong. *Journal of hazardous materials* 138 73.
[8] Zhou L, Xu X, Ding G, Miao Q 2003 The Correlation Factors And Pollution Forecast Model For PM2.5 Concentration In Beijing Area. *Acta Meteorologica Sinica* 61 761
[9] Zhao CX, Wang YQ, Wang YJ, Zhang HL, ZHao BQ 2014 Temporal and Spatial Distribution of PM2.5 and PM10 Pollution Status and the Correlation of Particulate Matters and Meteorological Factors During Winter and Spring in Beijing Environmental Science. *Envir. Sci.* 35 418.
[10] Yue Y, Wang X, Zhang M, Cao W, Zhou Y, Chen S, Zhu Y 2016 Air quality condition in Wuhan and its relationship to meteorological factors *Torrential Rain and Disasters* 35 271.
[11] Yu JZ, Tung JWT, Wu AWM, Lau AKH, Louie PKK, Fung JCH 2004 Abundance and seasonal characteristics of elemental and organic carbon in Hong Kong PM10. *Atmos. Environ.* 38 1511.
[12] Louie P, Watson J, Chow J, Chen A, Sin D, Lau A 2005 Seasonal characteristics and regional transport of PM in Hong Kong. *Atmos. Environ.*

Component, Disperse and Morphological Composition of Ambient Air Dust Contamination in the Zones of Mining-Processing Enterprises

S Y Zagorodnov, A A Kokoulina and S V Klein

FBSI "Federal Scientific Center for Medical and Preventive Health Risk Management Technologies", Perm, Russia

Abstract. The article presents the results of complex studies of dust emissions of the mining and processing complex. The main technological processes and the sources operation that produce intense dust emissions are determined. It has been proved that dust emissions of the investigated enterprise contain fine dust dangerous for human health. The results of the studies allowed us to detect precisely the dispersed composition of dust emissions, with the separation of PM10 and PM2.5 fractions, the chemical composition of the dust, and the shape of the particles. Thus, operating with obtained data on dispersed composition of the dust emissions and the specified sedimentation coefficient, we were able to calculate the dispersion of all solid particles and separately the PM10 and PM2.5 at the location of the enterprise. The dust exposure of the population at the targeted zones has been also determined. The obtained concentration values were used for assess health risk level to population living at the border of sanitary protection zone of the enterprise. Due to the obtained results, the enterprise was provided with recommendations on the inclusion of fine particles PM 10 and PM 2.5 in the production control program. In the case of increasing its production capacity was recommended to introduce the environmental measures for the reduction of emission of the finely dispersed fractions PM10 and PM 2.5.

1. Introduction

Dusty mixtures, emitted by industrial enterprises of the mining and processing complex, contain chemical components – heavy metal oxides, nitrates, sulfates, soot, etc. Particular danger to human health is represented by dust emissions containing fine particles up to 10 microns. This danger is becoming more and more obvious. From 1990 to 2010, 3.1 million people died of the causes associated with PM2.5 particles. Another figure: particles PM2.5 reduce the expected life span by an average of 8.6 months. Totally 3% of deaths from cardiovascular and respiratory system diseases and 5% of deaths from lung cancer are associated with PM2.5 [1, 2].

The next important point in the study of fine particles is their volatility. Dust emissions from industrial enterprises contain particles of various sizes: large, medium, small, while large dust particles freely accumulate near the plant. Small particles PM 2.5 accumulate more slowly, it is more difficult for them to overcome the resistance of the environment and accumulate on the ground. For the smallest particles, the resistance is also caused by Brownian motion. Being in suspension, fine particles are carried over long distances, creating dangerous concentrations in the territory of the population residence [3].

2. Purpose of the Study

Due to the fact that numerous negative effects of PM10 and PM2.5 fine dusts on human health have been proved by numerous studies [4, 5], as well as the possibility of adsorption of toxic components, including heavy metals, on dust particles, the goal of the present study was to evaluate quantitatively the component and dispersed composition of dust pollution of the ambient air in the zones of location of the primary magnesium production complex, as well as to describe the features of dust particles' morphology.

3. Object of the Study

The research object became a large mining and processing enterprise of the Perm Krai - the leader of the magnesium and rare metal industries of the Russian Federation, which accounts for almost 100% of REE, niobium and tantalum, more than 60% of commercial magnesium in the country [6].

4. Methods of Research

Sampling was carried out directly on the main sources of dust emission during the operation of process equipment, at a distance as close as possible to the source of dust emission. The distance depended on the specifics of the technological process. During the selection, filter materials were used to maximize the preservation of the fractional composition of the dust. The dispersion composition of the emissions was evaluated using a Microtrac S3500 laser particle analyzer (covered particle size range varied from 20 nm to 2000 μm); particle morphology and component composition were determined by electron microscopy using a high-resolution scanning microscope (magnification degree from 5 to 300 000 times) with a X-ray fluorescent attachment S3400N "HITACHI". The chemical composition was identified by X-ray phase analysis of the samples using an XRD-700 X-ray diffractometer "Shimadzu". To determine the level of dust pollution and to establish the emitting zones of influence of fine dust particles PM 10 and PM 2.5 and their comparison with the zones of influence of dust particle emissions without taking into account the dispersion, the model simulation of the dispersion of particulate pollutans' emissions from the sources of the enterprise has been conducted. Dispersion calculations were carried out using a unified program for the calculation of air pollution, implementing the provisions of OND-86 [7]. The calculations were verified by instrumental research.

5. Results of the Study

The obtained results of the research have demonstrated, that the disperse structure of dusts in emissions of various technological processes is heterogeneous. The results of instrumental studies of the determination of the different fractions portion (%) in the dust particles are shown in Table 1.

Table 1. Dispersed composition of dust emissions of the investigated sources of the mining and processing enterprise

№ s/p	Point of selection (Technological operation)	The content of particles of normalized fractions		
		$PM_{2.5}$,%	PM_{10},%	More than 10 micron
1	2	3	4	5
1	Unloading of finished products	10.77	26.03	73.97
2	Loading (steam hydrolysis section)	56.70	76.43	23.57
3	Unloading (steam hydrolysis section)	13.69	16.69	83.31
4	Unloading of finished products (precipitation of REM carbonates)	20.34	85.60	14.40
5	Preparation of soda solution	1.63	9.88	90.12
6	Mg refining, alloy preparation	0.00	0.56	99.44
7	Grinding of flux	2.14	19.96	80.04
8	Grinding of sulfur	3.33	18.79	81.21

IOP Conf. Series: Earth and Environmental Science **78** (2017) 012004 doi:10.1088/1755-1315/78/1/012004

9	Casting of metal on a foundry conveyor	4.55	15.53	84.47
10	Draining of SHES from the scrap crucible into the process box	0.00	0.514	99.49
11	Filter cleaning (chlorine after electrolysis)	11.66	38.63	61.37
12	Lifting and filling of supply bins	0.00	0.00	100.00
13	Unloading into the hopper (lime for neutralization of chlorine) lime after roasting	5.97	30.13	69.87
14	Loading of CaF2	3.35	30.98	69.02
15	Distillation of carnallite dust from in / ovens	9.75	44.48	55.52
16	Loading of chlorators with charge	9.02	28.99	71.01
17	Transportation of carnallite to the KS furnace	15.31	27.60	72.40
18	Transportation of carnallite from the KS furnace	8.04	41.05	58.95

The fraction of particles up to 2.5 μm in size (PM 2.5) inclusive was from 0.00 to 56.70%; Particles of less than 10 μm in size (PM 10) inclusive - 0.00 - 85.60%; Particles with sizes more than 10 microns - 14.40 -100.00%. Under microscopic study, the presence of nano-sized particles was established.

The distribution of the dispersed composition of dust emissions of the dust particles studied in the ejection structure is presented in the form of histograms in figure 1.

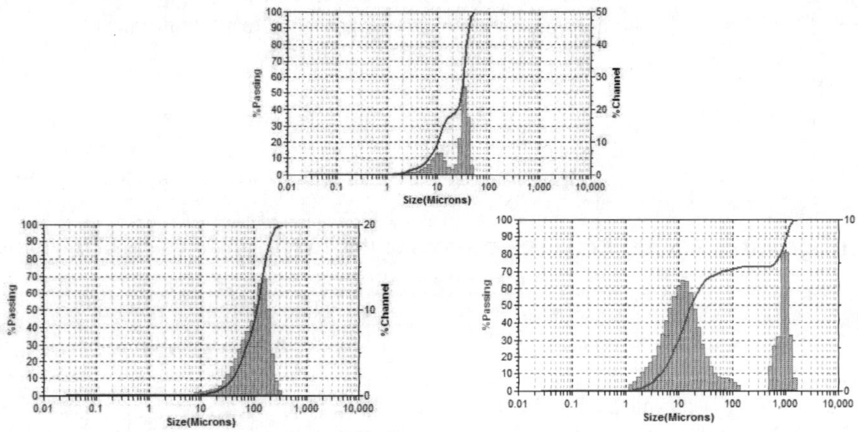

Figure 1. Percentage of suspended particles in industrial emissions.

In the course of the study, the component composition of the dust emissions of the selected samples was studied. The main share of the component composition of emissions consists of impurities distinctive for this production: magnesium, potassium, calcium, chlorine, silicon and metal oxides: iron, aluminum, lead, copper, manganese, zinc, chromium. Figure 2 shows the spectrograms of the emission components from the operation of the technological equipment of the investigated enterprise.

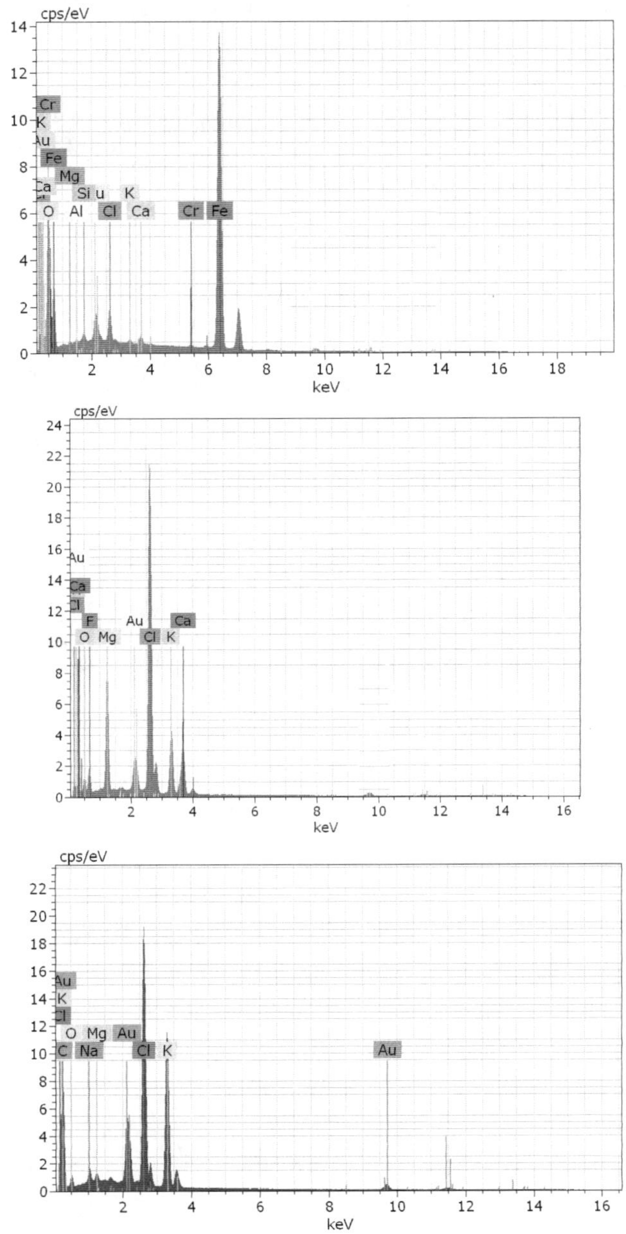

Figure 2. Spectrograms of the component composition of dust emissions of the investigated enterprises.

The studied morphological features of the selected samples allowed us to identify the shapes of fine particles emitted into the atmospheric air. It is determined that the dust particles have a rounded, irregular, cubic and composite shape. Images of dust particles of different shapes are shown in figure 3. Based on the obtained results of the disperse composition and morphological features, model calculations of the dispersion of dust emissions were carried out and zones of distribution of fine dust PM 10 and PM 2.5 over the territory were established and compared with the zones of influence of dust particle emissions without allowance for dispersion.

As initial data, the inventory data the pollutant sources of the investigated enterprise were used. The calculation took into account the emissions of solid pollutants and finely dispersed dusts PM10 and

PM 2.5. For each release of solid pollutants and finely dispersed dusts PM10 and PM 2.5, the settling coefficient (sedimentation) was refined [8]. Calculation of dispersion was carried out at the areas on the border of the sanitary protection zone of the enterprise and on a regular grid covering the territory of the enterprise and the outskirts including the territory of the population habitation.

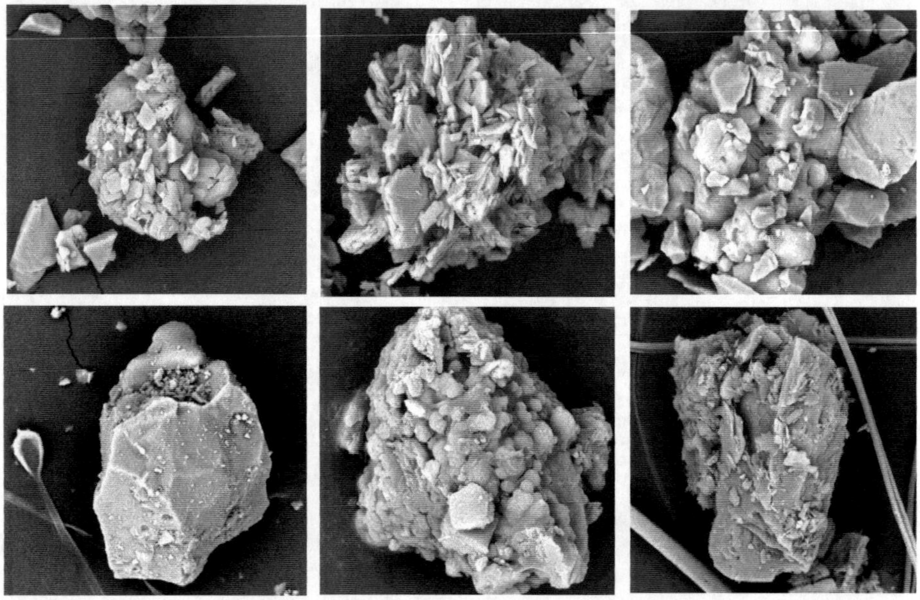

Figure 3. Forms of fine particles of the mining and processing complex.

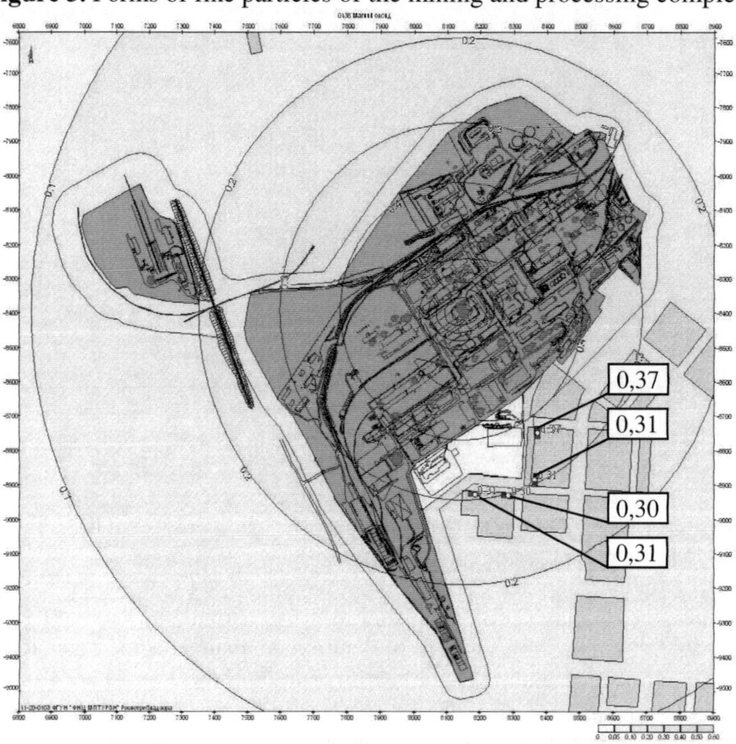

Figure 4. Isolation of dispersion of magnesium oxide (without taking into account the dispersed composition).

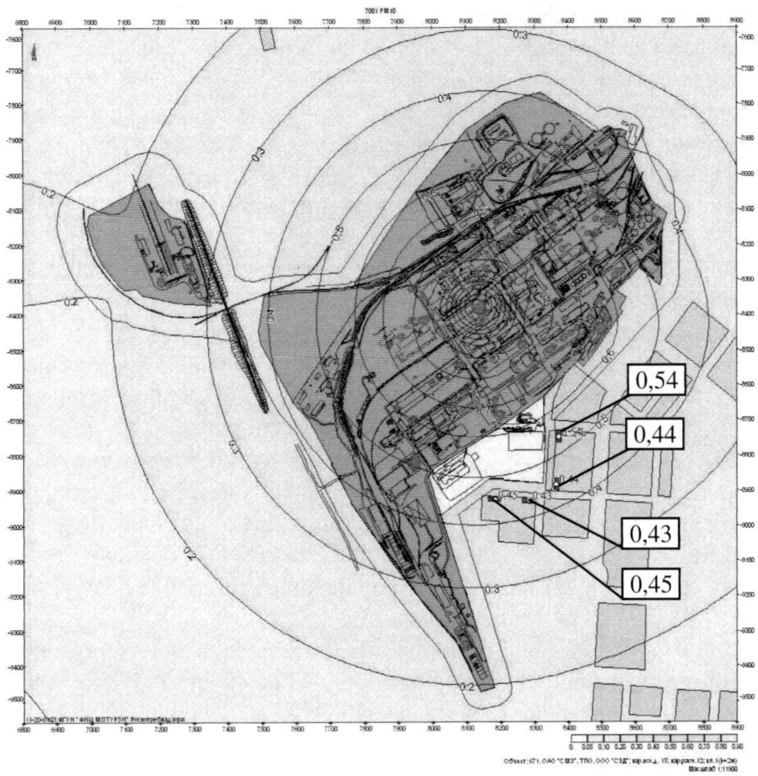

Figure 5. Isolation of dispersion of PM 10.

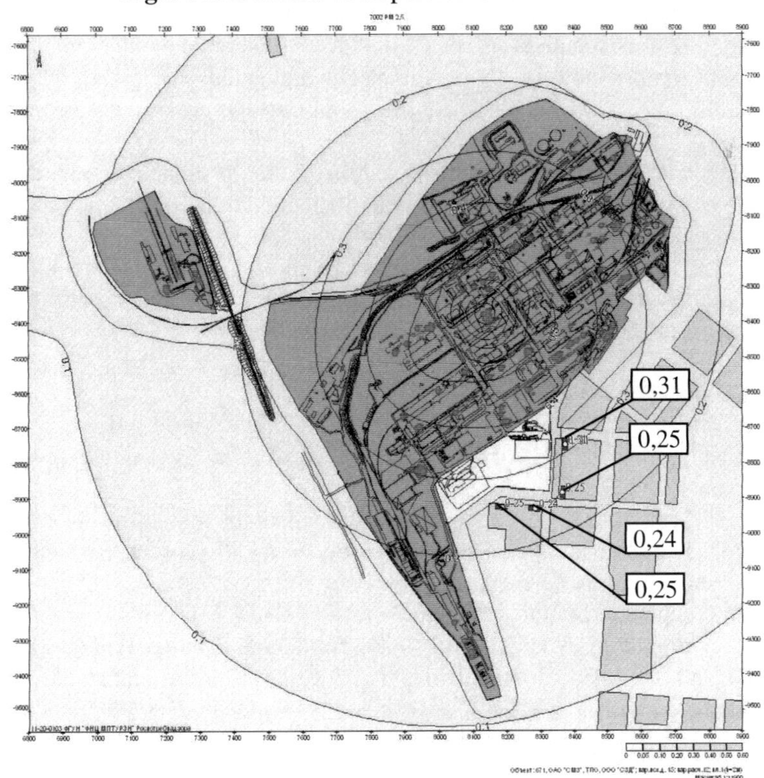

Figure 6. Isolation of dispersion of PM 2.5.

Calculation of the dispersion of dust emissions without taking into account the dispersed composition, according to the substances specific for the given production, established concentrations in the population's living areas for magnesium oxide - 0.37 MPC m, for calcium oxide - 0.14 MPC m. Taking into account the dust emissions of finely divided fractions separately for PM10 and PM2.5, the calculated concentrations of finely divided fractions at the population residential areas were established for RM 10 - 0.54 MPC m, for RM 2.5 - 0.31 MPC m. , Which indicates a higher level of air pollution. More than 200 people will live in the zone of the established exposition at the moment, including about 20 children.

In general, the situation should be assessed as a safe one, the normative quality of atmospheric air is observed at the border of residential areas.

Areas of distribution of solid dust emissions without taking into account the dispersed composition and fine particles in the form of isolines of dispersion are presented in figures 4-6.

Analysis of indicators of the health risk of the population [9] at the population residential areas determined that the maximum value of the coefficient characterizing the acute inhalation effect for fine particles is PM10 - 1.07 HQ, PM 2,5 – 0,77 HQ, which allows to assume risks of the formation of respiratory functions disorders in the population permanently residing near the enterprise, as well as other deviations in the state of health, characteristic for the action of finely dispersed particles.

The availability of up-to-date data on the dispersed composition of dust emissions made it possible to establish exposure levels to such hazardous components as PM 10 and PM 2.5, as well as to more accurately assess the health risks for the population.

Based on the results of the conducted studies, the enterprise was recommended to include in the production control program finely dispersed particles of PM 10 and PM 2.5 at the areas of population residence. In case of increasing production capacity, the introduction of environmental measures to reduce emissions of finely dispersed fractions of PM10 and PM 2.5.

6. Conclusions

- It is established that the technological processes of the mining and processing complex are accompanied by intense dust emission;
- it is proved that majority of the sources of the investigated enterprise contain emissions of fine particles dangerous for human health. Dust emissions into the atmosphere contain PM10 to 85.60%, PM2.5 to 56.70%;
- according to the study, it is clear that the probability of the presence of hazardous small particles in the air in the zones of influence of production facilities is quite high. The population living in the area of the enterprise is under the influence of dust factor;
- availability of actual information on dust emissions - component composition, dispersed composition, particle shapes, allow more accurate assessment of the exposure level at specified areas and apply the methodology for assessing the health risk for the population.

7. List of References

[1] Health effects of particulate matter, World Health Organization 2013.
[2] Cardiovascular Disease and Air Pollution. A report by the Committee on the Medical Effects of Air Pollutants 2006.
[3] Williams M.M.R. And Loyalka S.K. Aerosol science: theory and practice. - 1991, 446 p.
[4] Wiesław A. Jedrychowski, Frederica P. Perera, John D. Spengler, Elzbieta Mroz, Laura Stigter, Elżbieta Flak, Renata Majewska, Maria Klimaszewska-Rembiasz, Ryszard Jacek // J. Hyg. Intrauterine exposure to fine particulate matter as a risk factor for increased susceptibility to acute broncho-pulmonary infections in early childhood. Environ Health. 2013.
[5] Wilson R. and Spengler J. Particles in Our Air: Concentrations and Health Effects. - Cambridge, MA: Distributed by Harvard University Press, 1996.
[6] http: // sms.rf/index/istorija_predprijatija/0-19.
[7] OND-86 The procedure for calculating the concentrations in the air of harmful substances contained in the emissions of enterprises. All-Union normative document / GGO Voeikov. - L., 1987; P.64.

[8] May IV, Max A.A., Zagorodnov S.Yu., Chigvintsev V.M. Methodical approaches to the calculation of the sedimentation rate of various dust fractions for the purpose of assessing the population exposure to fine particles. // News of the Samara Scientific Center of the Russian Academy of Sciences. - 2012. - Vol. 14, No. 5 (3). 971- 975 pp.
[9] P 2.1.10.1920-04. A Guide for Assessing the Health Risks of the Population under Exposure to Chemicals Contaminating the Environment. Moscow, 2004.

IOP Conf. Series: Earth and Environmental Science **78** (2017) 012005 doi:10.1088/1755-1315/78/1/012005

Spatial Variation and Assessment of Heavy Metal and Radioactive Risk in Farmland around a Retired Uranium Mine

Jie LIANG, Chen-hao SHI, Guang-ming ZENG, Min-zhou ZHONG and Yu-jie YUAN

College of Environmental Science and Engineering, Hunan University, Changsha, P.R.China

Abstract. In recent years, heavy metal contamination in the environment has been attracted worldwide attention due to their toxicity, persistence,extensive sources and non-biodegradable properties. We herein investigate variation trend and risk of heavy metal and radiation distribution in the former mine stope, former mineral ore stockyard, and mine road with surface soils of a retired uranium mine in the mid-south of China. The mean concentrations (mg/kg) of Pb,Cd,Cu,Zn,As,Hg,Cr,Mn,Ni,U, and ^{232}Th were analyzed according to the corresponding background values in Hunan, China. The Geo-accumulation index (I_{geo}) were used for the assessment of pollution level of heavy metals and the radioactive elements of U and ^{232}Th. Then, Pollution load index (*PLI*) and GIS techniquewere integrated to assess spatial distribution of heavy metal contamination and radioactive contamination. Results confirmed that three areas in the retired uranium mine was a primary source of pollution, which showed anthropogenic origin mainly from agricultural runoff, hydrometallurgy from chemical industries, radioactive tailings, and electroplating industriesfinally drained into Zishui River and Xiangjiang River. Based on the actual situation, some suggestions were put forward for the treatment of the retired uranium mine in conclusion.

1. Introduction

In recent years, heavy metal and radioactive elements contamination in the environment has been attracted worldwide attention due to their toxicity, persistence, extensive sources and non-biodegradable properties [1]. The agricultural soil around the mining area, and the quantities are more than the carrying capacity of heavy metals. The food chain can serious impact one the health of the human body, and ultimately cause health risks to the human body[4]. The result of the principal components (PCs) and the factor analyze (FA) showed that the anthropogenic activities are primary sources of soil contamination with heavy metals and radioactive elements.

Combined with multivariate statistical analysis and geographic information system (GIS) technology[7], we investigated the three regions around the retired uranium mine of heavy metal and radioactive pollution sources in the mid-south China. Specifically, the heavy metal and radioactive element was sampled and analyzed in the former mine stope, former mineral ore stockyard and mine road. Therefore, the following three aims are considered in this study.(1) analyzing pollution status of nine heavy metals (Pb,Cd,Cu,Zn,As,Hg,Cr,Mn, and Ni) and two natural radioactivity (U and ^{232}Th), (2) using potential geo-accumulation index(I_{geo}), and combined pollution load index (*PLI*) and GIS to assess spatial variation of heavy metal contamination and (3) using the Pearson's coefficient to analysis the correlation with factor analysis (FA) between the heavy metal elements and radioactive elements.

2. Materials and Methods

2.1. Study Area

The uranium mine, retired in 1996, is located in Hunan Province, middle and southern part of China. The type of ore deposit belongs to the hard rock type uranium deposit in the little hilly area with the elevation of 200-300m. The average annual temperature is 18 - 20°C, and the mean annual precipitation is 1286-1372mm. The polluted surface soil contains heavy metal elements, such as Pb, Zn, Cd, and so on, and radioactive elements of U and ^{232}Th. The corroded soil form farmland flows into the river, lies in the south of the study area.

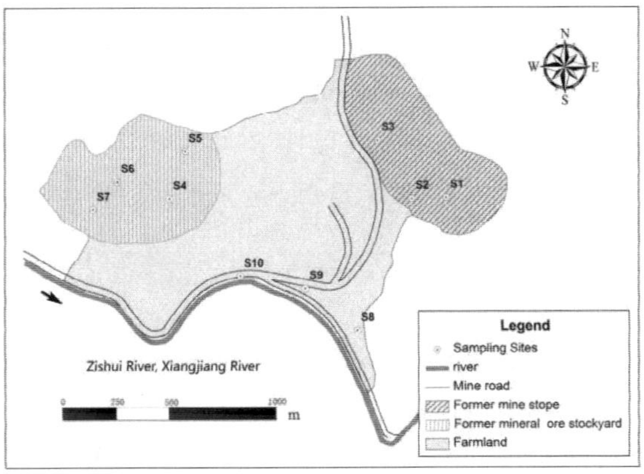

Figure 1. Study area and sampling sites of surface soils in farmland along a retired uranium mine.

2.2. Sampling

In Spring season of 2016, a total of ten sites were collected in the farmland around the retired (Figure 1). In the former mine stope, former mineral ore stockyard and mine road collected nearly 60 soil samples. In each samples sites, the former mine stope, former mineral ore stockyard and mine road collected respectively 24, 16 and 18 soil samples. S1, S2, and S3 were located in mine stope. S4, S5, S6 and S7 were located in mineral ore stockyard. S8, S9, and S10 were located in mine road. The polyethylene bags were packed and sent back to the laboratory for cold storage.

2.3. Analytical methods and quality control

The sample was dried up in an oven at 100°C for 24h, until constant weightsieved with 40 mesh sieve for homogenization, and homogenized and separated into two parts. A sample of 0.5g was weighed and transferred to airtight Teflon vessels and added by HNO_3, HF, $HClO_4$ which were digested with a mixture. The other sample of 328.0g was weighted and sealed in airtight polyethylene cylindrical container, and then placed for 30 days to reach secular equilibrium before further gamma-ray measurement. Inductively coupled plasma-massspectrometry (ICP-MAS), followed by the Chinese DZ/T0223-2001, was used to analyze Pb, Cd,Cu,Zn,Cr,Ni. Atomic fluorescence spectrometry (AFS) was used to analyze As, Hg, Mn.Gamma spectrometry (GS) was used to analyze ^{232}Th,and U, follow by the Chinese GB/T 11743-2013.

In the analysis process, the quality control of the soil composition analysis standard materials (GBW07401, Institute of Geophysical and Geochemical Exploration) was added to meet the quality control requirements. The analytical precision was conducted with the repetitive rate of 10%.[8] The results met the quality control requirement of The Technical Specification for Soil Environmental Monitoring (HJ/T 166-2004) and Determination of radionuclides in soil by gamma spectrometry (GB/T 11743-2013).

2.4. Geo-accumulation index(I_{geo})

I_{geo} index was introduced by Muller(1969).[9] It is a geochemical criterion to evaluate heavy metal pollution and radioactive pollution in uranium mining area. The equation is also considered the background value change which caused by the natural diagenesis. The I_{geo} was calculated by the equation [10]:

$$I_{geo} = \log_2(C_n / 1.5B_n) \quad (1)$$

Where C_n is the measured concentration of element (n), mg/kg and B_n is the geochemical background value of the element (n), mg/kg. In the equation, average values were used, and 1.5 is the factor used for lithologic variations of trace metals. According to the soil background values in Hunan(1988), background values of heavy metal and radioactive contents in a uranium mine were used as B_n for calculation. The level classification of I_{geo} is defined as unpolluted(<0),unpolluted to moderately polluted (0-1),moderately polluted (1-2),moderately to strongly polluted (2-3), strongly polluted (3-4),strongly to super strongly polluted (4-5) and super-strongly polluted(>5).

Table 1. Summary statistics of heavy metal and radioactive concentrations in the retired uranium mine, compared with guide values of China(mg/kg, dry weight).

	Pb	Cd	Cu	Zn	As	Hg	Cr	Mn	Ni	^{232}Th	U
Former mine stope											
Min	173.8	1.78	38.33	280.54	18.12	165	29.63	300.54	3.8	18.82	16.23
Max	23190	22.54	228.89	4173	226.2	1881.76	60.66	572.21	18.26	84.62	217.22
Mean	3568.	10.8	119.6	2150.	57.8	839.	47.4	435.1	12.3	27.89	101.

	57	3	6	49	6	69	5	4	5		46
S.D.[a]	5854.71	6.062	57.97	1250.32	45.37	554.34	8.05	74.31	3.4	16.79	53.31
CV[b] (%)	1.64	0.56	0.48	0.58	0.78	0.66	0.17	0.17	0.28	0.6	0.53
Former mineral ore stockyard											
Min	145.71	0.63	21.8	184.88	15.89	113.43	31.21	333.4	4.17	22.29	14.12
Max	17370	10.26	157.21	2548.96	92.15	509.51	64.74	504.66	17.08	60.01	527.77
Mean	2479.48	2.69	45.57	655.03	37.94	225	47.1	429.73	11.95	30.47	91.27
S.D.[a]	4948.66	2.73	41.5	720.8	19.99	107.76	8.96	45.12	3.98	10.18	126.48
CV[b] (%)	2.0	1.01	0.91	1.1	0.53	0.48	0.19	0.11	0.33	0.33	1.39
Mine Road											
Min	50.7	0.11	12.84	63.98	15.29	62.1	29.07	310.31	8.43	20.81	6.88
Max	1323	12.4	58.58	1976	62.85	1082.74	56.39	494.46	17.18	28.12	43.75
Mean	370.7	2.79	30.37	515.28	43.37	300.57	43.81	398.88	11.63	23.63	22.2
S.D.[a]	383.85	3.4	13.42	567.46	13.84	297.78	6.69	40.76	2.24	1.99	10.95
CV[b] (%)	1.04	1.22	0.44	1.1	0.32	0.99	0.15	0.1	0.19	0.08	0.49
Mean of 3 fields	2275.69	6.09	71.51	1230.47	47.87	502.81	46.22	422.39	12.01	27.28	74.05
Background[c]	27	0.085	26	94	14	0.096	68	441	32	17	4.2

[a]S.D.: standard deviation
[b]CV: coefficient of variation
[c]Background: According to the Soil background values in Hunan (1988)

2.5. Pollution load index (PLI)

Pollution load index (*PLI*) is a kind of evaluation method proposed by TOMLINSON[12] in the classification of heavy metal pollution levels[13]. Contamination factor(*CF*) is the ratio of measured concentration by background value. For the entire sampling site, *PLI* has been determined as the nth root of the product of the *nCF* [14]:

$$CF = C_n / B_n \quad (2)$$

$$PLI = (CF_1 * CF_2 * CF_3 * ... * CF_n)^{1/n} \quad (3)$$

The evaluation steps include the calculation of the highest pollution coefficient and the calculation of the pollution load index, and to provide a straightforward and comparative method for evaluating the level of heavy metal and radioactive contamination. The level classification of *PLI* is defined asunpolluted(<1),moderately polluted (1-2),strongly polluted(2-3) and super-strongly polluted(>3).

2.6. *Multivariate statistical and GIS-based analyses*

In this study, we used the Pearson correlation coefficient to reflect the linear relationship between variables. All the statistical analyses were performed using SPSS 18.0 for Windows. ArcGIS 10.0 software was used to analyze the spatial variations of heavy metal and radioactive elements [16].

3. Results and discussion

3.1. *Soil contamination characteristics*

The mean contents of Pb,Cd,Cu,Zn,As,Hg,Cr,Mn,Ni,^{232}Th,and U in three areas were presented in Table 1,along with background values in Hunan, China. In surface soils of a uranium mine.Heavy metal elements and radioactive elements were higher than the background values, such as Pb, Cd,Cu, Zn, As, Hg and U. Pb,Cd,Cu,Zn,Hg and U in the former mine stope and the former mineral ore stockyard were approximately twice to forth as high as that of mine road.

The pollution level of Pb,Cd,Cu,Zn,Hg and U has increased the degree of pollution during the three areas studiedin Figures 2 and Table 2.Hg was more than two times in the super-strongly pollution level. The same level of Cd pollution is not optimistic, and the serious is only less than Hg. Zn and U were increased quickly with all elements. Pb was in the moderately level. According to I_{geo}, average pollution level decreased following the order Hg>Cd>Pb>U>Zn>As>Cu>^{232}Th>Mn>Cr>Ni.

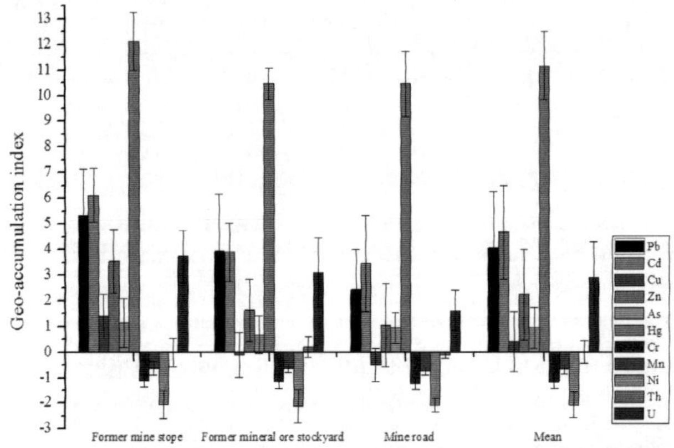

Figure 2. Geo-accumulation index(I_{geo})of heavy metals and radiationsin a retired uranium mine for three areas.

Table 2. The level classification of Geo-accumulation index (I_{geo}) of heavy metals and radiations in a retired uranium mine for three areas.

	Former mine stope	Former mineral ore stockyard	Mine road	Mean
Pb	Super-strongly polluted	Strongly polluted	Moderately to strongly polluted	Strongly to super-strongly polluted
Cd	Super-strongly polluted	Strongly polluted	Strongly polluted	Strongly to super-strongly polluted
Cu	Moderately polluted	Unpolluted	Unpolluted	Unpolluted to moderately polluted
Zn	Strongly polluted	Moderately polluted	Moderately polluted	Moderately to strongly polluted
As	Moderately polluted	Unpolluted to moderately polluted	Unpolluted to moderately polluted	Unpolluted to moderately polluted
Hg	Super-strongly polluted	Super-strongly polluted	Super-strongly polluted	Super-strongly polluted
Cr	Unpolluted	Unpolluted	Unpolluted	Unpolluted
Mn	Unpolluted	Unpolluted	Unpolluted	Unpolluted
Ni	Unpolluted	Unpolluted	Unpolluted	Unpolluted
^{232}Th	Unpolluted	Unpolluted to moderately polluted	Unpolluted	Unpolluted to moderately polluted
U	Strongly polluted	Strongly polluted	Moderately polluted	Moderately to strongly polluted

3.2. Spatial risk assessment of heavy metal and radioactive elements

Spatial risk map of heavy metal in a retired uranium mine was shown in Figures 3. To facilitate the analysis, with different colors to represent the various points in the extreme heavy pollution conditions over the multiple samples. In the former mine stope, it might be affected by the residue soil and hydrometallurgical mill which left the radioactive contamination and chemical pollution. In the former mineral ore stockyard, S4,S5,S6, were also the more severe contamination area, which was nearly the uranium deposit cave and hydrometallurgical mill. There are many ditches, and finally, import to Zishui River. Infected by rain and slag mucks, the heavy metal and radioactive were polluted around the farmlands. S7 is lower than S4,S5, and S6, due to it was far from the ditches. In mine road, S8, S9, and S10,the area were located nearly the battery factory and chemical plant. The heavy metal pollution was more serious than radioactive contamination. Similarly affected by the ditches, the pollution levels were only below the mine yard.

3.3. Correlation and factor analysis

Correlation analysis of heavy metal and radioactive in soils were shown in Table 3. In the former mine stope, a strong positive correlation between Pb,Cd,Cu,As,Cr,Mn,Ni,^{232}Thand U($P \leq 0.01$) was presented. In the former mineral ore stockyard, a strong positive correlation between Cd,As,Hg,Cr,Mn,Ni,^{232}Th, and U is shown. In mine road, both all elements were found a definite positive. The ^{232}Th and U were also found a strong positive correlation in three fields. They were randomly distributed in the surface soil and considered in association with other heavy metal elements.

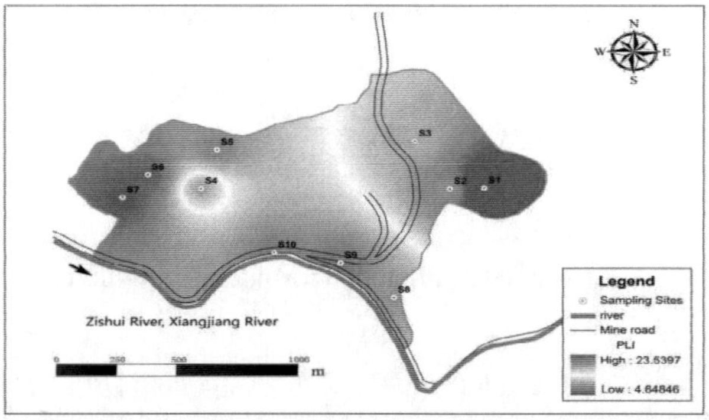

Figure 3. Spatial distribution maps using pollution load index(*PLI*) for heavy metal and radioactive risk in a uranium mine.

Table 3. Pearson correlation matrix for heavy metal and radioactive concentration in a retired uranium mine.

	Pb	Cd	Cu	Zn	As	Hg	Cr	Mn	Ni	^{232}Th	U
Former mine stope											
Pb	1										
Cd	0.274^{a}	1									
Cu	0.412^{a}	0.6^{a}	1								
Zn	0.419	0.962^{a}	0.647^{a}	1							
As	0.199^{a}	0.503^{a}	0.115^{a}	0.502^{a}	1						
Hg	0.083^{b}	0.541^{a}	0.441^{a}	0.493^{a}	0.399^{a}	1					
Cr	-0.579^{a}	-0.442^{a}	-0.244^{a}	-0.515^{a}	-0.408	-0.544^{a}	1				
Mn	-0.113^{b}	-0.124^{a}	-0.381^{a}	-0.158^{a}	0.267^{a}	-0.17^{a}	0.256^{a}	1			
Ni	-0.79^{a}	-0.212	-0.299^{a}	-0.287^{a}	-0.249^{a}	-0.118^{a}	0.573^{a}	0.232^{a}	1		
^{232}Th	-0.232	0.97	0.09	0.299	0.26^{a}	0.10	-0.06^{a}	-0.29	0.332	1	

	[a]	3[a]	4[a]	[a]		2[a]		5[a]	[a]		
U	0.456[a]	0.754[a]	0.862	0.781[a]	0.295[a]	0.407[a]	-0.304[a]	-0.112[a]	-0.295[a]	0.332[a]	1
Former mineral ore stockyard											
Pb	1										
Cd	0.918	1									
Cu	0.988	0.961[a]	1								
Zn	0.951	0.994[a]	0.979[a]	1							
As	0.089	0.191[a]	0.11	0.176[a]	1						
Hg	0.784	0.94[a]	0.848[a]	0.918[b]	0.124[a]	1					
Cr	-0.494	-0.475[a]	-0.496	-0.471[a]	-0.23[b]	-0.346[a]	1				
Mn	-0.504[b]	-0.515[a]	-0.521[a]	-0.5[b]	-0.155[a]	-0.291[a]	0.763[a]	1			
Ni	-0.693	-0.627[a]	-0.684[a]	-0.635[a]	-0.295[a]	-0.431[a]	0.92[a]	0.814[a]	1		
^{232}Th	0.915	0.797[a]	0.886	0.843[a]	-0.06	0.666[a]	-0.4[a]	-0.355[a]	-0.542[a]	1	
U	0.958	0.8[a]	0.913	0.85[a]	0.107	0.629[a]	-0.49	-0.439[a]	-0.681[b]	0.93	1
Mine road											
Pb	1										
Cd	0.397[a]	1									
Cu	0.901[a]	0.676[a]	1								
Zn	0.485	0.99[a]	0.736[a]	1							
As	-0.029[a]	0.205[a]	-0.065[a]	0.153[a]	1						
Hg	0.034	0.202[a]	0.251[a]	0.181	-0.441[a]	1					
Cr	-0.725[a]	-0.336[a]	-0.667[a]	-0.407[a]	-0.272	0.343[a]	1				
Mn	-0.64	-0.097[a]	-0.418[a]	-0.256	-0.295[a]	0.152[b]	0.557[a]	1			
Ni	-0.457[a]	-0.147[a]	-0.266[a]	-0.191[a]	-0.253[a]	0.175[a]	0.475[a]	0.775[a]	1		

^{232}Th	-0.166^a	-0.058^a	-0.09^b	-0.091^a	-0.244^a	0.598^a	0.567^a	0.278^a	0.028^a	1	
U	0.824^a	0.525^a	0.884	0.56^a	0.033^a	0.225^a	-0.567^a	-0.549^a	-0.298^a	-0.112^a	1

^{a}Correlation is significant at the 0.01 level(2-tailed)
^{b}Correlation is significant at the 0.05 level(2-tailed)

Loading of the principal components (PCs) for three areas were shown in Figure 4a-4c. In the formelmine stope, PC1 (53.1% of the total variance) was dominated mainly by Zn, Cd, and Hg, and the origin could be from chemical industries. PC2 (35.6% of the total variance) was dominated primarily by Cu, As, and Pb, the origin could be from agricultural runoff. PC3(17.2% of the total variance) was dominated mainly by U and ^{232}Th. The origin could be from radioactive tails. In the former mineral orestockyard , PC1 explained 41.6% of the total variance and dominated by Zn, and Hg, it might be chemical industries. PC2 explained 39.2% of the total variance and dominated by Pb; it might be electroplated industries. PC3 explained 16.4% of the total variance and dominated by U and ^{232}Th; it might be radioactive tailings. In mine road, PC1 (43.5% of the total variance) was dominated mainly by Hg and As; the origin could be from agricultural runoff. PC2 explained 41.7% of the total variance and dominated by Pb, Cd, and Zn; it might be chemical industries. PC3 accounted for 12.9% with the strong loading of U and ^{232}Th; it was similarly to mineral ore stockyard and mine stope.

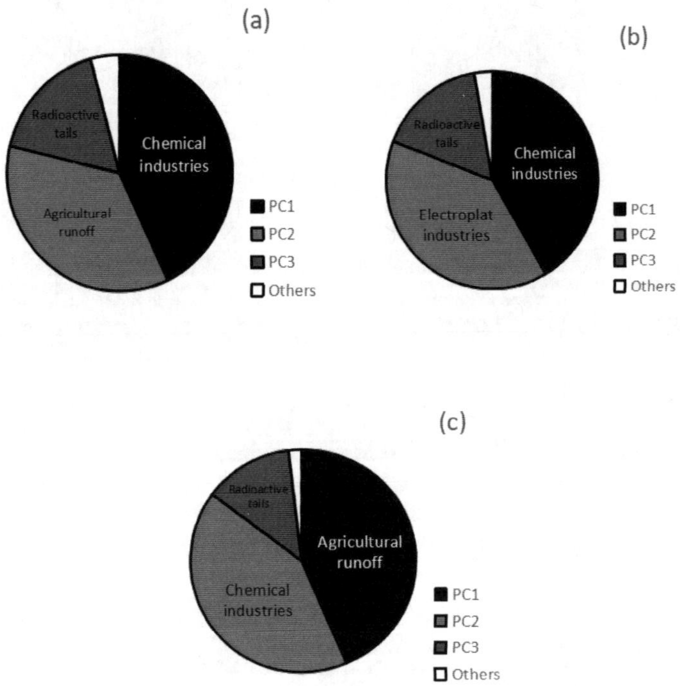

Figure 4. Loadings of the PCs for heavy metals and radiations: (a) in the former mine stope, (b) in the former mineral ore stockyard and (c) in mine road.

4. Conclusion

In this article, three-fields longitudinal assessment and analyses were conducted in the present study in a uranium mine during dry seasons. After the investigation, we carried out nine heavy metals and two radioactive elements and revealed their pollution condition. Pb,Cd,Hg,Zn, and U were in potentially high risk with considerable loadings and increased contents. Especially Hg and Cd were in super-strongly pollution status. Hg was more than two times. Due to the growth of agricultural runoff, chemical industries from hydrometallurgy, radioactive tailings, and electroplate industries, they were the factors that lead to heavy metal and radioactive risk. Spatial distribution analyses indicated the former mine stope and the former mineral ore stockyard were the biggest sources of pollution, and the a key to retire governance. A strong positive correlation between Pb,Cd,As,Cr,Mn,Ni,^{232}Th and U($P \leq 0.01$) is found. Finally, we suggested taking measures to reduce the risk in a retired uranium mine.

Due to the soil-plant system is a source of bio-energy, the plants can clean up the potential risk of the heavy metal pollution [17] .Restore soil fertility is a way of accelerate ecological succession, such as different grass species planting, legumes and native species rotating [18]. Natural and artificially constructed wetlands get helpful to control the elements of heavy metal in acid mine water, and avoid spreading the heavy metal pollution in surface and subsurface water [19]. The vegetables with radionuclides can not to feed the livestock in the uranium mine, and they are growing on the sludge and the stream wetland [20]. Avail the susceptibility of the long half-life of radionuclides, such as post-reduction minerals to reoxidation and remobilisation via microbial metabolism or abiotic mechanisms, and it reduces the radioactivity in the environmental remediation [21]. The chemical treatment get helpful to control the radionuclides in tailings, such as $CaO/CaCO_3$ and $BaCl_2$, and cover with clay materials and other soils [22].

This study indicated that the environmental pollution problem of the retired uranium ore can not be ignored and got the first line information of the risk assessment of the domestic retired uranium mine.The obtained data can be not only used as a reference data for monitoring possible pollution in future but also can be applied to radioactive site remediation technology, phytoremediation technology, and cover experiment.

5. Acknowledgments

This work is funded by the National Natural Science Foundation of China (51479072, 51679082,51521006, 51579094) the New Century Excellent Researcher Award Program (NCET-08-0181) from the Ministry of Education of China and the Program for Changjiang Scholars and Innovative Research Team in University (IRT-13R17).

6. References

[1] MAO Long-jiang, MO Duo-wen, FU Qiang, YANG Jing-hong, JIA Yao-feng. Multivariate analysis of heavy metals in surface sediments from lower reaches of the Xiangjiang River, southern China [J]. Environmental Earth Sciences, 2013, 69(3); 765-771.

[2] ZHOU Jian, MA Dong-sheng, PANJia-yong, NIEWen-ming, WU Kai. Application of multivariate statistical approach to identify heavy metal sources in sediment and waters: a case study in Yangzhong, China [J]. Environmental Geology, 2008, 54(2); 373–380.

[3] ZHU Hui-na, YUAN, Xing-zhong, ZENGGuang-ming, JIANG Min, LIANGJie, ZHANG Chang, YIN Juan, HUANG Hua-jun, LIUZhi-feng, JIANG Hong-wei. Ecological risk assessment of heavy metals in sediments of Xiawan Port based on modified potential ecological risk index [J].Transactions of Nonferrous Metals Society China, 2012, 22(6), 1470–1477.

[4] RENZONI A, ZINO F, FRANCHI E. Mercury Levels along the Food Chain and Risk for Exposed Populations [J]. Environmental Research, 1998, 77(2), 68-72.

[5] DUDKA S, MILLER W P. Accumulation of potentially toxic elements in plants and their transfer to human food chain [J]. Journal of Environmental Science and Health, Part B, 1999, 34(4), 681-708.

[6] PERALTA-VIDEA J R, LOPEZ M L, NARAYAN M, SAUPE G, GARDEA-TORRESDEY J. The biochemistry of environmental heavy metal uptake by plants: Implications for the food chain [J]. The internal Journal of Biochemistry & Cell Biology, 2009, 41(8-9), 1665-1677.

[7] LIANG Jie, LIU Jia-yu, YUAN Xing-zhong, ZENG Guang-ming, LAI Xu, LI Xiao-dong, WU Hai-peng, YUAN Yu-jie, LI Fei. Spatial and temporal variation of heavy metal risk and source in sediments of Dongting Lake wetland, mid- south China [J]. Journal of Environmental Science and Health Part A, 2015, 50(1), 100-108.

[8] ZHANG Man-yin, CHUI Li-juan, SHENG Lian-xi, WANG Yi-fei. Distribution and enrichment of heavy metals among sediments, water body and plants in Hengshuihu Wetland of Northern China [J]. Ecological Engineering, 2009, 35(4), 563-569.

[9] HARLIBALA, Hu Bi-tao,WANGCheng-guo,GERILEMANDAHU, Hu Xiao, ZHANG Shuai, BAO Shan-hu, LI Yu-hong. Assessment of radioactive materials and heavy metals in the surface soil around uranium mining area of Tongliao, China [J]. Ecotoxiclogy and Environmental Safety, 2016, 130(8), 185-192.

[10] PRAVEENA S M, AHMED A, RADOJEVIC M, ABDULLAH H, ARIS A Z. Heavy Metals in Mangrove Surface Sediment of Mengkabong Lagoon, Sabah: Multivariate and Geo-Accumulation Index Approaches [J]. Bulletin of Environmental Contamination and Toxicology, 2008, 81(1), 52-56.

[11] LIU Ping-hui, WEI Chang-shuai, ZHANG Shu-mei, ZHU Chuang-ming, XIE Shu-rong. Assessment on Radioactive Uranium Contamination of Paddy Soil in Uranium Mine at Southeast China by ICP-MS [J]. Asian Journal of Chemistry, 2015, 27(3), 1049-1052.

[12] TOMLINSON D L, WILSON J G, HARRIS C R, JEFFREY D W. Problems in the assessment of heavy-metal levels in estuaries and the formation of a pollution index [J]. Environmental Evaluation, 1980, 33(1), 566-575.

[13] AHMED F, BIBI H M, FUKUSHIMA T, SETO K, ISHIGA H. Recent sedimentary environment of coastal lagoon in southwestern Japan:evidence from major and trace elements [J]. Environmental Monitoring and Assessment, 2011, 173(1), 167-180.

[14] ANGULO E. The Tomlinson Pollution Load Index applied to heavy metal, 'Mussel-Watch' data: a useful index to assess coastal pollution [J]. Science of The Total Environment, 1996, 187(1), 19-56.

[15] SURESH G, RAMASAMY V, MEENAKSHISUNDARAM V, VENKATACHALAPATHY R, PONNUSAMY V. Influence of mineralogical and heavymetal composition on natural radionuclide concentrations in the river sediments [J]. Applied Radiation and Isotopes, 2011, 69(10), 1466-1474.

[16] LU G Y, WONG D W. An adaptive inverse-distance weighting spatial interpolation technique [J]. Computers & Geosciences, 2008, 34(9), 1044-1055.

[17] CHENG Shui-ping. Heavy metal pollution in China: Origin, pattern and control [J]. Environmental Science and Pollution Research, 2003, 10(3), 192-198.

[18] WONG M H. Ecological restoration of mine degraded soils, with emphasis on metal contaminated soils [J]. Chemosphere, 2003, 50(6), 775-780.

[19] SHEORAN A S, SHEORAN V. Heavy metal removal mechanism of acid mine drainage in wetlands: A critical review [J]. Minerals Engineering, 2006, 19(2), 105-106.

[20] CARVALHO F P, OLIVEIRA J M, MALTA M. Radionuclides in plants growing on sludge and water from uranium mine water treatment [J]. Ecological Engineering, 2011, 37(7), 1058-1063.

[21] LLOYD J R, RENSHAW J C. Bioremediation of radioactive waste: radionuclide - microbe interactions in laboratory and field-scale studies [J]. Current Opinion in Biotechnology, 2005, 16(3), 254-260.

[22] FERNANDES H M, FRANKLIN M R, VEIGA L H S, FREITAS P, GOMIERO L A. Management of Uranium Mill Tailing: Geochemical Processes and Radiological Risk Assessment [J]. Journal of Environmental Radioactivity, 1996, 30(1), 69-95.

Research on Model and Related Parameters of Supercritical CO_2 Injection into Depleted Reservoir

Ma Pinghua[1] **and He Jun**[1]

1 College of Vehicle and Energy, Yanshan University, Qinhuangdao, Hebei, 066004, China

Abstract. On the basis of the research on CO_2 geological storage and enhanced oil recovery(EOR) technology, a random porosity distribution model which conforms to logarithmic normal distribution was adopted in order to describe the heterogeneous characteristics of pore structure. On this basis, the two-phase flow model of CO_2-formation water was established to describe the displacement process. Through the simulation of CO_2 injection into depleted reservoir, it confirmed that injection point pressure was associated with the temperature and depth of the formation except heterogeneity. But the saturation distribution was greatly influenced by formation heterogeneity and depth. Thus, the space utilization of the injection layer reduced gradually with the depth increasing. The related research is important for CO_2 storage, migration and evolution in depleted reservoir.

1. Introduction

With the rapid development of modern industrial economy, the increasing oil consumption results in many oil blocks losing economic exploitation value and becoming depleted reservoirs[1]. Meanwhile, as the main component which discharged from the burning of fossil fuels, carbon dioxide (CO_2) is considered to be one of the greenhouse gases that may contribute most to global warming on the earth[2-4]. Disposal of CO_2 from stationary sources into subsurface structures has been suggested as a possible means for reducing CO_2 emissions into the atmosphere. The geological storage has been studied as one of the effective ways of CO_2 emission reduction[5-6]. As one of the ideal storage place, depleted reservoir is selected for carbon dioxide sequestration, and its oil recovery also can be improved by CO_2-flooding during CO_2 injection. However, the oilfield has injection wells, production wells and equipments, which can greatly reduce the carbon dioxide capture and recycling costs in the early stage of storage. Depleted reservoir has a natural closed space and conditions, carbon dioxide in depleted reservoir can be mainly storaged by dissolving into the residual oil and formation water sealed in the formation. In addition, carbon dioxide can react with the high valence cations of the formation water and sealed.

Depleted reservoir which have lost its economic exploitation value, is mostly in a state of high water cut. Porous media of these reservoirs is mainly composed of formation water (or injected water) with highly scattered residual oil distributing[7]. Considering the oil-water contact, two-phase flow of CO_2 and water is studied firstly during CO_2 being injected into the depleted reservoir. The pore structure of formation changed a lot due to the long-time water flooding, physical and chemical reaction among CO_2 and water and rocks[8]. Thus, the original permeability and porosity of the reservoir are not accurate any more. Given the characterization of porosity can be characterized by average value, the random simulation method is used to describe the heterogeneous characteristics of porosity.

A random porosity distribution model which conforms to logarithmic normal distribution is adopted in order to describe the heterogeneous characteristics of pore structure which had been changed in the injection layer. On this basis, two-phase flow model of CO_2-formation water is established to describe the displacement process. Considering the operation parameters influence on the characteristics of CO_2 injecting into depleted reservoirs, the effects were analyzed through the numerical simulation. The impacts of porosity heterogeneity, injection depth and injection pressure were mainly discussed in this paper, in order to provide a theoretical basis for CO_2 sequestration and migration in the depleted reservoir. Meanwhile, the results can be referenced for the further studying on WAG.

1.1. Porosity Distribution Model

The logarithmic normal distribution function is used to describe the average porosity of the injection layer. As the porosity distribution model is built, heterogeneous pore structure characteristics of CO_2 injection layer can be studied.

Logarithmic normal distribution of probability density function can be expressed as follows:

$$f(x)=\frac{1}{x\sigma\sqrt{2\pi}}\exp\left[-\frac{(\ln x-\theta)^2}{2\sigma^2}\right] \tag{1}$$

Among them, the parameter θ and σ are the mean and standard deviation of the variables logarithm respectively. Its mathematical expectation is the average porosity of injection layer which can be expressed as: $\phi = e^{\theta+\sigma^2/2}$. Its Variance is the variance of the porosity which can be expressed as: $\varepsilon = \left(e^{\sigma^2}-1\right)e^{2\theta+\sigma^2}$, The relationship between them can be expressed as:

$$\theta = \ln\left(\frac{\phi^2}{\sqrt{\varepsilon+\phi^2}}\right) \tag{2}$$

$$\sigma = \sqrt{\ln\left(\frac{\varepsilon+\phi^2}{\phi^2}\right)} \tag{3}$$

Rock porosity is 0.1~0.4 in commonly, considering the different reservoir conditions of Depleted reservoirs which has been developed by water flooding for long time, 0.25 is selected as average porosity and 0.01 as its variance in this study. Thus, the parameter θ and σ can be calculated by formula (2) and (3): θ=-1.4605, σ =0.3853.

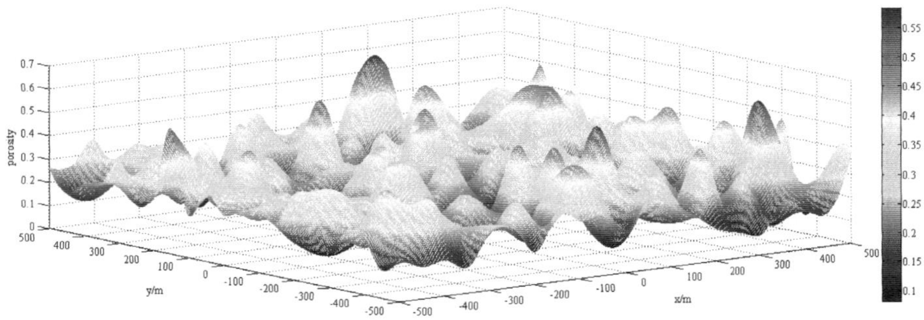

Figure 1. Porosity distribution.

The porosity distribution model can be solved through MATLAB7.0 as follows:

①Function meshgrid() is used for meshing the regional of calculation model, to generate the compute nodes;

②In the region of two-dimensional, the function lognrnd () is used to generate the porosity array which satisfy θ and σ in the grid nodes and interpolation method is used on the outside space. calculating the two-dimensional area so as to realize the porosity distribution of random logarithmic normal distribution. As shown in Figure 1. And the pore variance can be 0 if Pore formation is homogeneous.

2. Mathematical Model and Computation Conditions

The critical pressure of CO_2 is 7.38 MPa, and the critical temperature is 31.04°C. Under the supercritical state, CO_2 will become a supercritical fluid, which has as big as liquid density, but at the same time is as high as the gas dispersion coefficient and low viscosity[9].

2.1. Research Area

The model of two phase flow in porous media was employed to develop a simulation code of CO_2 injection. The two phases are assumed to be CO_2 fluid as non-wetting phase and groundwater as wetting phase. The CO_2 density was calculated by the Peng-Robinson equation, and CO_2 viscosity was calculated by the Jossi-Stiel-Thodos equation. The formulation of the numerical model and theoretic explanation is as follows.

Established mathematical model based on the following assumptions: (1) CO_2 is in the supercritical state.(2) Ignore the solubility of supercritical CO_2 in formation water;(3) Ignore the capillary force in reservoir pore;(4)The geothermal gradient is only considered in supercritical CO_2 density and the influence of viscosity;(5) Fluid and rocks are incompressible , the fluid flow conform to darcy law.

2.2. The Establishment of the Two-Phase Fluid Continuity Equation and the Equation of Motion

The mass conservation equation is as follows:

$$\frac{\partial(\phi S_w \rho_w)}{\partial t} + \nabla \cdot (\rho_w v_w) = q_w \qquad (4)$$

$$\frac{\partial(\phi S_n \rho_n)}{\partial t} + \nabla \cdot (\rho_n v_n) = q_n \qquad (5)$$

where the subscript w for the water phase and n for the CO_2 phase, ϕ is the porosity, S the saturation, q the density, t time and v the Darcy flux velocity.

Darcy's law extended to two phase flow can be written for each phase as

$$v_w = -\frac{Kk_{rw}}{\mu_w}(\nabla p_w - \rho_w g) \qquad (6)$$

$$v_n = -\frac{Kk_{rn}}{\mu_n}(\nabla p_n - \rho_n g) \qquad (7)$$

where K is the absolute permeability, k_{rw}, k_{rn} the relative permeability, μ_w, μ_n the viscosity, p_w, p_n the pressure and g the acceleration due to gravity.

2.3. Equation of State for Two Phase

The multiphase system in porous media is characterized by parameters such as the capillary pressure, relative permeability and saturation relationship. The pressure difference across the interface between two immiscible fluids was neglected under the assumption that the injection pressure would be much higher than the capillary pressure ($p_w = p_n$), so the capillary pressure p_c =0.

The two phase saturation is satisfied as follows:

$$S_w + S_n = 1 \qquad (8)$$

According to the van Genuchten model, the relationships between the relative permeabilities and the effective saturation are given by

$$k_{rw} = S_e^{\alpha}\left[1-\left(1-S_e^{1/\psi}\right)^{\psi}\right] \qquad (9)$$

$$k_{rn} = (1-S_e)^{\gamma}\left(1-S_e^{1/\psi}\right)^{2\psi} \qquad (10)$$

where S_e is the effective saturation, which can be described as follows:

$$S_e = \frac{S_w - S_{wr}}{1 - S_{wr} - S_{nr}} \qquad (11)$$

Where S_{wr} and S_{nr} are the water and CO_2 residual saturation respectively, α, ψ and γ are parameters that are determined by the shape of the pores. According to the literature' typical values for the parameters α, ψ and γ are 1/2, 1/3 and 0.77, respectively.

In order to describe the changes of density and viscosity of supercritical CO_2 caused by the changes of temperature and pressure in the injection layer, the Peng-Robinson equation is used to calculate the density of supercritical CO_2, and the Jossi-Stiel-Thodos equation is used to calculate the viscosity of supercritical CO_2 in this paper.

2.4. Boundary Conditions

For the study area and the establishment of model equations, the initial conditions and boundary conditions are as follows:

Initial conditions: when the t=0, $p=f(z)$, S_w=0.9, S_n=0.1

The injection point is located in the centre of the calculation region, and CO_2 saturation is 1 in the injection point while simulating the CO_2 injection. In the calculation region, the initial saturation of CO_2 is 0.1, except for the injection point.

Boundary conditions: In the calculation, the pressure and initial saturation are given by the pressure gradient at the boundary, which is the first kind of boundary condition:

p=f(z)，x=±500；z=±500

S_w=0.9，S_n=0.1，x=±500；z=±500

$q=q_0$，x=0；z=0

2.5. Solving Model

The commercial simulation software of COMSOL Multiphysics is used in order to solve the mathematical model. Coupled with the Peng-Robinson equation, the Jossi-Stiel-Thodos viscosity equation and Porosity model conforms to logarithmic normal distribution. Table 1 is the part of the parameter values used in the model[5].

Table 1. Parameter values in the model.

attribute paramete	parameter values
Calculation area	1000m×1000m
Permeability K/10^{-3}μm^2	100
Average porosity of injection layer ϕ	0.2
Formation water density ρ_w/kg·m^{-3}	1.0×10^3
Formation water viscosity μ_w/mPa·s	0.283
CO_2 viscosity μ_n/mPa·s	by Jossi-Stiel-Thodos
CO_2 density ρ_n/kg·m^{-3}	by Peng-Robinson
CO_2 residual saturation S_{nr}	0.05
Residual water saturation S_{wr}	0.05
Pressure gradient /MPa·km^{-1}	10.5
Geothermal gradient /K·km^{-1}	45
Surface temperature /K	293
Injection rate q_0/kg·s^{-1}	5

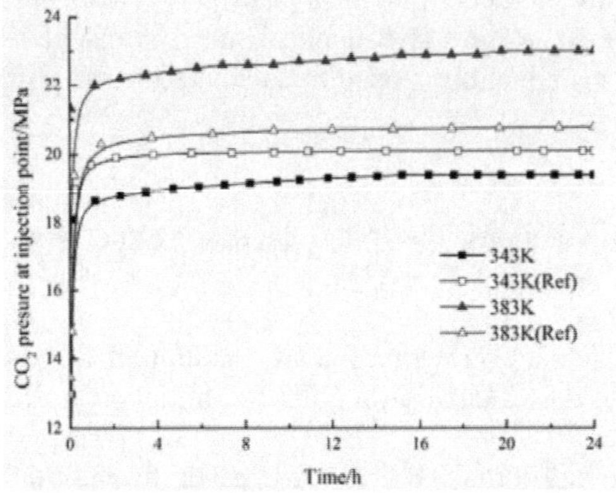

Figure 2. Comparison of pressure at the injection point of two different viscosity models .

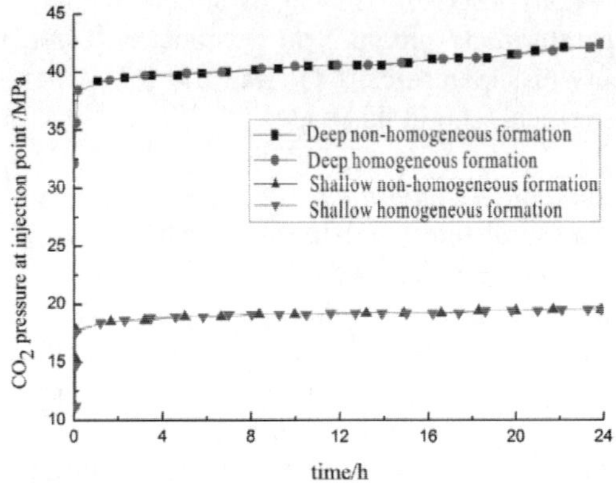

Figure 3. changes of injection point pressure within 24 hours.

Through the simulation, the change of injection point pressure at different temperatures was obtained (Figure 2). At first, the results of this model are compared with the results of K Sasaki et al. In the calculation process, the initial conditions and relevant parameters used in this paper are most of the same as that used by Sasaki K(Table 1)[5], and the difference is that the viscosity of supercritical carbon dioxide, which is used by Sasaki K et al is a fixed value of 0.216×10^{-1}mPa·s,and in this paper, the Jossi-Stiel-Thodos equation is used to characterize the viscosity which adapted to the formation temperature and pressure, the calculated viscosity range is 0.206×10^{-1}mPa·s to 0.394×10^{-1}mPa·s.

As Figure 2 shows that, at the same temperature, the viscosity values obtained by different viscosity models have obvious effects on the injection pressure of carbon dioxide. When the injection temperature is 343K (70℃), the injection point pressure variation curve which corresponding to the viscosity value calculated by the equation of Jossi-Stiel-Thodos is lower than the injection pressure which corresponding to the fixed viscosity, and when the injection temperature is 383k (90℃), the result is just the opposite. Therefore, in the formation of large changes in temperature and pressure, in order to more accurately characterize the influence of the actual viscosity of carbon dioxide on injection point pressure, using viscosity equation which is adapted to local temperature and pressure formation to characterize is more in line with the actual situation of formation.

Under a certain temperature, the pressure of the injection point will increase gradually and tends to be stable over time. The higher the reservoir temperature, the greater the pressure of the injection point, which is due to the density of carbon dioxide decreases with the increase of temperature. The higher the temperature is, the longer the injection pressure is stable.

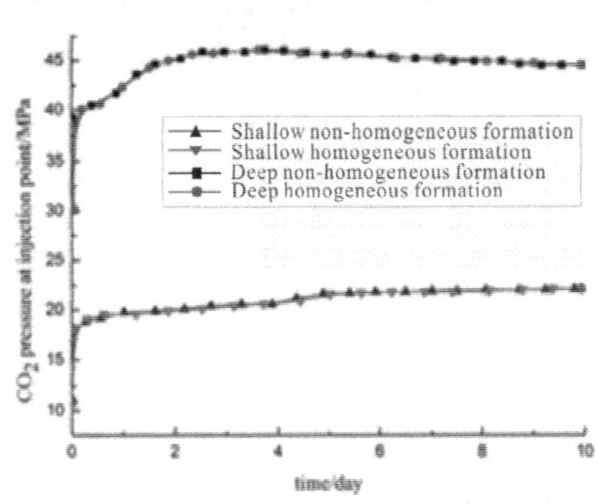

Figure 4. changes of injection point pressure in 10 days.

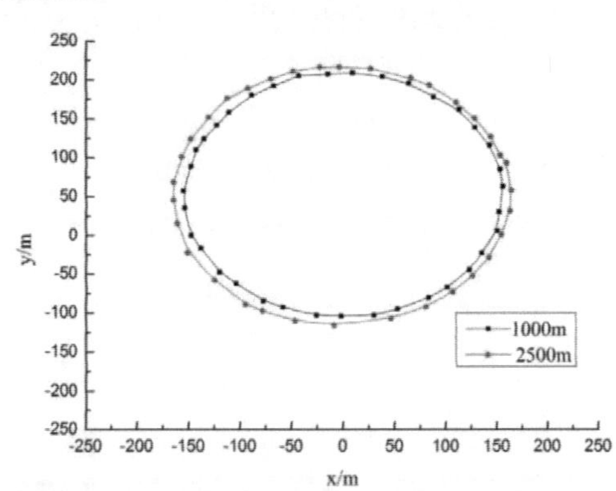

Figure 5. The edge distribution of CO_2 in homogeneous strata of different depth.

In previous studies, scholars have mostly assumed that the formation is homogeneous strata, and not made a note for whether the result can accurately express the change of the actual formation pressure. To this end, the injection point pressure of homogeneous and heterogeneous strata in different depth are simulated and compared in the study(Figure 3 and Figure 4). From the figure we can see that the heterogeneity of the formation has little effect on the pressure of the injection point. In both shallow and deep strata, the injection pressure is related to the formation depth, which is not related to the heterogeneity of the formation. The deeper the formation is, the greater the pressure of the injection point is, and the pressure increases rapidly with time and tends to stabilize to a relatively stable value.
Due to the influence of geothermal gradient and pressure gradient, the flow characteristics of supercritical CO_2 in different depth strata will also change. Figure 5 is the saturation distribution characteristics of CO_2 in the injected layer after injecting 10 days into homogeneous strata of 1000m and 2500m depth with the injection rate of 5kg/s respectively. The closed line in the figure is the edge of CO_2 which is sealed in the strata. As can be seen, under the same conditions of injection rate and time, with the increase of the depth, the outer edge of CO_2 has increased. Showing that the space utilization of the injection layer reduced gradually with the depth increasing. Under the same volume, the amount of CO_2 stored in the deep reservoir is smaller than that in the shallow reservoir, but the injection pressure is larger than that of the latter. Considering the long-term storage of sealing, the deep reservoir still has a certain practical application value and advantage. At the same time, because of the density difference between formation water and supercritical CO_2, the influence of the buoyancy effect makes the CO_2 distribution shift to the surface direction (Figure 5). This phenomenon also exists in the homogeneous formation, but saturation distribution of CO_2 are greatly influenced by the heterogeneity of porosity. In the larger porosity region, the flow ability is stronger, and then produce irregular distribution characteristics, the simulation results will be further discussed in the later study.

3. Conclusions

(1)The CO_2 injection point pressure is little affected by the heterogeneity of porosity, and the injection point pressure is mainly related to the depth of the injection layer, the average porosity, the injection rate and formation temperature.
(2)The viscosity of CO_2 has obvious influence on the pressure of the injection point. Therefore , it is more consistent with the actual situation of the formation to be characterized by the viscosity equation adapted to the local temperature and pressure.
(3)Due to the existence of the density difference between the supercritical CO_2 and the formation water, the buoyancy effect makes the CO_2 to the surface of the earth's surface, which is great significance to grasp the migration and evolution of CO_2 after storage.
(4)At the same injection condition, affected by the geothermal gradient and pressure gradient, temperature and pressure in the deep strata are larger, which results in CO_2 viscosity decreases, and flow ability enhanced. In the deep injection layer, the space utilization rate of CO_2 is relatively low, and the injection point pressure greatly. But from the point of view of the effect of long-term storage, deep formation has certain advantages.

4. References

[1] Zhangliang, Wangshu, et al. Assessment of CO_2 EOR and its geo-storage potential in mature oil reservoirs[J]. 2009 *Petroleum exploration and development*. 36 737~42.
[2] Zhoudi. 2005 CO_2 geological storage - new subject of geology[J]. *Progress in natural science.* 15 782~7.
[3] Sunshu. 2006 Geological Problemsof CO_2 Underground Storage and Its Significance on Mitigating Climate Change[J]. *China Basic Science*. 8 17~22.
[4] Qiangwei, Li Yilian, Wen Dongguang, et al. 2006 Advances and Problems of Geological Disposal of Greenhouse Gases[J]. *Geological Science and Technology Information*. 2006 25 83~8.
[5] Sasaki K, Fujii T, Niibori Y, et al. 2008 Numerical Simulation of Supercritical CO_2 Injection into Subsurface Rock Masses[J]. *Energy Conversion and Management.* 49 54~61.

[6] Bachu S. 2008 CO_2 Storage in Geological Media: Role, Means, Status and Barriers to Deployment[J]. *Progress in Energy and Combustion Science.* 34 254~73.
[7] Zhao Xiaoliang, Liao Xinwei, Wang Wanfu, et al. 2013 Evaluative model of CO_2 geological sequestration and determination of key parameters[J]. *Special Oil & Gas Reservoirs.* 20 72~4.
[8] Xu T, Apps JA and Pruess K. 2003 Reactive geochemical transport simulation to study mineral trapping for CO_2 disposal in deep arenaceous formations. *J Geophys Res.* 108(B2) 3-1~3-13.
[9] Li Gensheng, Wang Haizhu, Shen Zhonghou, et al. 2013 Application investigations and prospects of supercritical carbon dioxide jet in petroleum engineering[J]. *Journal of China University of Petroleum.* 37 76~80.

Acknowledgements
This work was financially supported by "Colleges and universities in Hebei province science and technology research projects: Research on the Key Technology of Industrial Waste Gas Stored in Depleted Reservoir(QN2014048)".

Study on Photocatalytic Degradation of Endocrine Disrupting Compound

Bhagwan Pralhad Parihar, Smita Gupta, Mousumi Chakraborty*

Department of Chemical Engineering, Sardar Vallabhbhai National Institute of Technology Surat – 395 007, Gujarat, India
Email: mch@ched.svnit.ac.in, mousumi_chakra@yahoo.com

Abstract. Propylparaben (PP) is catagorized as endocrine disrupting compounds and is found to be present in urban wastewater comparatively at high concentrations. In the present work, propylparaben was degraded photo-catalytically by optimizing different process parameters such as initial concentration of propylparaben (25mgL^{-1} to 100 mgL^{-1}), pH of the feed phase and concentration of photocatalyst TiO_2 (50 mgL^{-1} to 200 mgL^{-1}). Finally PP degraded and converted to CO_2 and H_2O and the degradation was found to follow the first order kinetics.

1. Introduction

The most common techniques are to treat wastewater containing organic dyes, pigments and pharmaceutical compounds are classifiable into three main categories i.e. physical (adsorption, filtration, and flotation), chemical (coagulation, oxidation, reduction, electrolysis) and biological (aerobic, anaerobic degradation). However, due to the complexity and variety of organic compounds, it has become rather difficult to find a unique treatment procedure that covers the actual elimination of all toxic organic compounds. Particularly, biochemical oxidation suffers from significant limitations since most organic compounds found in the commercial market have been intentionally designed to struggle aerobic microbial degradation. Physical processes also have some limitation therefore the chemical process should be treated by alternative advanced processes. Photocatalytic degradation process is also one of the important advanced oxidation methods. This results in complete mineralization of wide range of the recalcitrant or hazardous organic compounds. Recently, it is found that parabens are frequently released in urban wastewater comparatively at high concentrations and, even with considerably removal of them using conventional treatment methods; they have been still identified in river water samples [1]. Hence, the U.S. Environmental Protection Agency has considered these compounds as emerging environmental pollutants [2]. From literature it is found that numerous articles are available on photocatalytic degradation of different endocrine disrupting compounds (Table 1) but very few have studied the photocatalytic degradation of PP using UV/TiO_2 system and that too at lower pollutant concentration.

So in this study, photocatalytic degradation of PP have been examined by varying the process parameters such as initial concentration of propylparaben (25 mgL^{-1} to 100 mgL^{-1}), pH of the feed phase and concentration of photocatalyst TiO_2 (50 mgL^{-1} to 200 mgL^{-1}) and kinetic study is also performed.

Table 1. Literature review on photocatalytic degradation of endocrine disrupting compounds

Author	pollutant	Oxidation condition	Remark

Alaton and Balcioglu (2001)	Reactive Black 5	H_2O_2=10Mm, H_2O_2/UV-A; pH=4;	COD 62%; TOC 75% in 2h[3]
Watanabe et al.(2002)	Bisphenol A	pH=3 ; TiO_2	90% degradation of BPA[4].
Katsumata et al.(2003)	Bisphenol A	pH=4; Fe(II)=$4*10^{-5}$ mol/L; H_2O_2 $4*10^{-4}$ mol/L	90% degradation in 36 h[5].
Li et al. (2007)	Bisphenol A	pH 3-4;Oxalate ; iron oxide	84% removal in 40 min[6].
Yamazaki et al.(2008)	4 tert-octyl phenol	$K_2S_2O_8=2*10^{-2}$mol dm^{-3}; TiO_2 peroxy disulphate	83.2% degradation in 6 h[7].
Bledzka et al.(2010)	n-butyl paraben &4 tert-octyl phenol	pH=7; H_2O_2=0.01M; H_2O_2/UV	90% degradation [8]
Yanlin et al.(2012)	4 tert-octyl phenol	pH=4.5; TiO_2 precursor sol 13.6% ; TiO2	90% degradation in 30 min[9].
Huange et al.(2012)	Bisphenol A	Neutral pH; Fe(III)- ethylene diamine N,N'-di succinic acid (EDDS) =0.1mM with 1mM of H_2O_2;	Complete degradation in oxygen saturated solution[10].
Yanlin et al.(2013)	4 tertoctyl phenol	pH=3.5; Fe(III)=$3*10^{-4}$ mol/L	80%degradation in 60 min[11].
Osarumwense et al. (2015)	Phenol in aqueous solution	Periwinkle shell ash (PSA) as photocatalyst.	90% degradation[12]
Hurtado et al. (2016)	phenol and 4-chlorophenol (4-CP)	coupled electro-oxidation/ozonation > electro-Fenton-like process > photo-Fenton process > heterogeneous photocatalysis	Comparison of different treatment processes[13]

2. Experimental Procedure

2.1. Materials and Method

Propylparaben (LOBA Chemie), silver nitrate (Sigma Aldrich) and TiO_2 (Finar) were purchased and used without further purification. All other chemicals HCl, NaOH, H_2O_2 used were of analytical grade. All the solutions were prepared in demineralized water. Calibration curve was plotted using 2 to 10 mg/L propylparaben (at 256 nm wavelength) solution on the UV-Vis spectrophotometer (HACH, Germany) to determine the rate of the degradation

2.2. Experimental Procedure

All experiments were performed in a batch reactor. The reactor was cylindrical with 150 ml volume and made of quartz glass for the transfer of the radiation. Irradiation was achieved by using UV lamp of 125 W (medium pressure lamp) which was immersed in the glass tube surrounded by cooling water jacket to maintain reactor temperature. The reaction chamber, between the reactor walls and UV lamp, was filled with the reaction mixture. Mixing was accomplished using air bubbler to keep the photocatalyst in suspension.

2.3. Mechanism of Photocatalytic Reaction

TiO_2 is used as photocatalyst to make the degradation process faster as TiO_2 is a semiconductor and act as an oxidizing agent. There are many oxidizing agents like hydrogen peroxide, magnesium peroxide, fluorine, potassium bromide etc. but in most of the research work, TiO_2 was used as photocatalyst because it produces of $OH^{\bullet}$ at a faster rate than other photocatalysts. When photo catalyst TiO_2 captures UV radiation from sunlight or illuminated light source (fluorescent lamps), the electron of the valance band of TiO_2 becomes excited. The excess energy of this excited electron of TiO_2 promotes the electron from valence band to the conduction band. So negative electron and positive hole are formed. The energy difference between valence band and conduction band is known as band gap energy. The positive hole of TiO_2 breaks apart the hydrogen molecule to form hydrogen gas and hydroxyl radical. The negative electron reacts with oxygen molecule to form super oxide anion. This cycle continues until sunlight or illuminated light source is available [14].
Photocatalytic reaction with TiO_2 photocatalyst occurs as follows [15]
On the TiO_2 oxidizing agent absorption of photon energy in the form of UV light occurs and then electron-hole pair formation takes placed.

$$TiO_2 + hv \rightarrow TiO_2^- + OH^{\bullet} \text{ (Or } TiO_2^+)$$

Produce extremely reactive but short lived hydroxyl radical ($OH^{\bullet}$) by hole trapping.

$$H^+ + H_2O \rightarrow OH^{\bullet} + H^+$$

Surface reaction i.e. oxidation reaction of the pollutant molecules under UV light to produce carbon dioxide and water.

$$OH^{\bullet} + O_2 + C_xO_yH_{(2x-2y+2)} \rightarrow xCO_2 + (x-y+1)\, H_2O$$

Radical recombination (energized holes and electrons can recombine)

$$TiO_2^- + OH^{\bullet} + H^+ \rightarrow TiO_2 + H_2O \text{ (recombination)}$$

3. Results and Discussions

3.1. Effect of pH on Feed Phase

The photocatalytic degradation of PP was carried out at different pH at 4, 6, 8 and 10. Maximum degradation was observed at pH 8. It was observed that the slight basic pH of feed phase was favourable for degradation of propylparaben while the acidic pH was not favourable. Figure 1, shows degradation rate decreased at acidic pH because, in acidic phase, the formation of $OH^{\bullet}$ radicals may be too fast. It is noticed that the optimum results was found out at pH 8. Thus, all the remaining experiments were conducted at pH 8 and further evaluation were considered on same basis.

3.2. Effect of TiO_2 loading

The photocatalytic degradation of PP was carried out at different concentration of TiO_2 (50-200 mg/L). Concentration of PP was kept 100 mg/L in the feed solution (Figure 2). Maximum degradation (58.39%) was observed at 50 mg/L of TiO_2. It might be due to UV light falls on the solution resulting into generation of sufficient number photons and electrons by UV light during the degradation reaction but thereafter with further increase in catalyst loading the degradation rate starts declining due to the screening effect of suspended TiO_2 present in the feed solution.

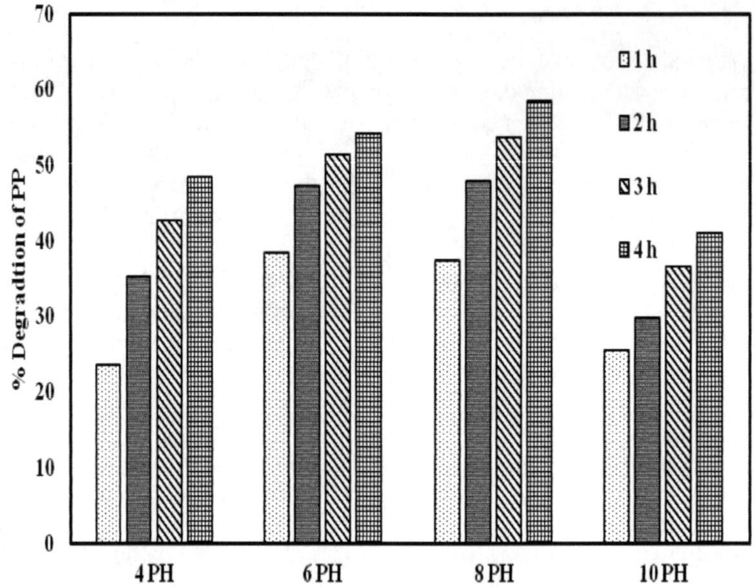

Figure 1. Effect of feed phase pH

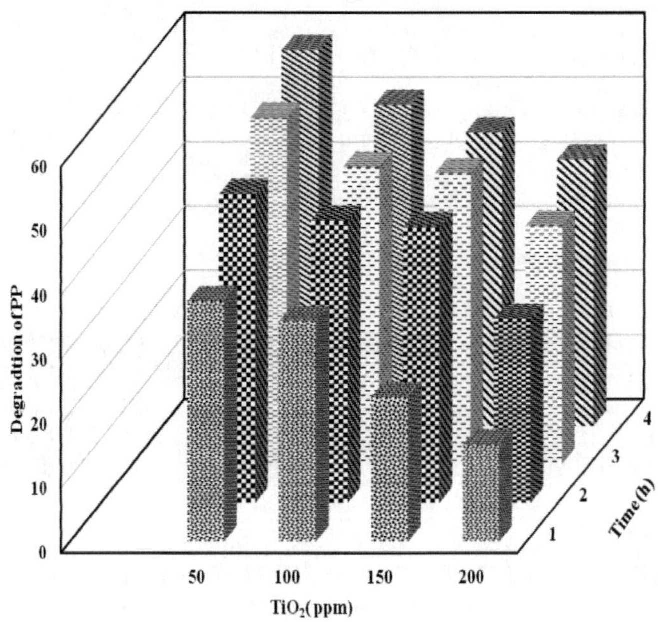

Figure 2. Effect of TiO_2 concentration

3.3. Effect of initial concentration of PP in feed phase

The degradation of PP was studied at 25, 50, 75 and 100 mg/L. It is found that as concentration of propylparaben solution increases, the rate of degradation decreases and 25 ppm of PP solution showed maximum 85.2 % degradation in 4h whereas degradation of 100 ppm solution was only 58.39%. Actually there was not enough dosage of TiO_2 to provide active radicals for the degradation of PP at higher initial concentration which also led to formation of more by- products, which might absorb some photons or consumed additional active radicals and decreased degradation rate.

3.4. Kinetic Study

Degradation kinetics was studied using 50 ppm TiO_2 concentration and 25 ppm PP solution as feed.

$$\ln (C/C_0) = -kt$$

Plot of ln (C/C_0) vs. time was plotted (not shown). It was observed that degradation followed first order kinetics and average rate constant value was found to be $k = 3.31\times10^{-4}\ s^{-1}$.

3.5. Energy Consumption

Major cost of degradation process is the electrical and chemical cost. For photocatalytic degradation energy consumption was calculated by this equation

Daily consumption (kWh) = Wattage x hours used per day/1000

The evaluation of energy consumption was calculated for 100% degradation. Time taken for 100% degradation was substituted in the above equation to calculate daily consumption of the power, which was found to be 2-3 kWh, quite cheaper than other advanced treatment technique

4. Conclusion

This study showed that photocatalytic degradation process was strongly pH dependent. Process efficiency of photocatalytic degradation also depends on feed concentration, dosing amount of the TiO_2 photocatalyst. Using 50 ppm TiO_2 photocatalyst, 25 ppm of PP showed maximum 85.2 % degradation in 4h whereas degradation of 100 ppm solution was only 58.39%. Comparing energy consumption with other advanced treatment technique, it was observed that photocatalytic degradation is more economical than other processes.

5. References

[1] I. Gonzalez-Marino, J. Quintana, I. Rodrıguez, R. Cela, Evaluation of the occurrence and biodegradation of parabens and halogenated by-products in wastewater by accurate-mass liquid chromatography-quadrupole-time-of-flight-mass-spectrometry, Water Res, 45 (2011) 6770-6780.

[2] Y. Lin, C. Ferronato, N. Deng, J. Chovelon, A. Catal., Study of benzylparaben photocatalytic degradation by TiO_2, Appl Catal Environ,104 (2011)353-360.

[3] I. Alaton, I. Balcioglu, Photochemical and heterogeneous photocatalytic degradation of waste vinylsulphone dyes a case study with hydrolysed Reactive Black 5, J. Photochem Photobiol A, 141 (2001) 247-254.

[4] N. Watanbe, S. Horikoshi, H. Kawabe, Y. Sugie, J. Zhao, H. Hidaka, Photo degradation mechanism for bisphenol A at the TiO_2/H_2O interfaces, Chemosphere, 52 (2003) 851-859.

[5] H. Katsumata, S. Kawabe, S. Kaneco, T. Suzuki, K. Ohta, Degradation of Bisphenol A in water by photo-Fenton reaction, J Photoch Photobio A Chem, 162 (2004) 297-305.

[6] F. Li, X. Li, X. Li, T. Liu, J. Dong, Heterogeneous photo degradation of bisphenol A with iron oxides and oxalate in aqueous solution, J. Colloid Interface Sci, 311 (2007) 481-490.

[7] S. Yamazaki, T. Mori, T. Katou, M. Sugihara, A. Saeki, T. Tanimura, Photocatalytic degradation of 4-tert-octylphenol in water and the effect of peroxydisulfate as additives, J Photochem Photobiol A Chem, 199 (2008) 330-335.

[8] D. Bledzka, D. Gryglik, M. Olak, J. Gebicki, J. Miller, Degradation of n-butyl paraben and 4-tert-octylphenol in H_2O_2/UV system, Radiant Phys Chem, 79 (2010) 409-416.

[9] Y. Wu, H. Yaun, X. Jiang, G. Wei, C. Li, W. Dong, Photocatalytic degradation of 4-tert-octylphenol in a spiral photo reactor system, J Environ Sci, 24 (2012) 1679-1685.

[10] W. Huange, M. Brigante, F. Wu, K. Hanna, G. Mailhot, Development of new homogeneous photo-Fenton process using Fe (III)-EDDS complexes. J Photochem Photobiol A Chem, 239 (2012) 17-23.
[11] Y. Wu, H. Yuan, G. Wei, S. Zhang, H. Li, W. Dong, Photo degradation of 4-tert-octylphenol in aqueous solution prompted by Fe(III). Environ Sci Pollut Res, 20 (2013) 3-9.
[12] J. O. Osarumwense, N. A. Amenaghawon, F. Aisien, Heterogeneous, photocatalytic degradation of phenol in aqueous suspension of periwinkle shell ash catalyst in the presence of UV from sunlight, J Eng Technol, 10 (2015) 1525-1539.
[13] L.Hurtado, D.Amado-Piña, G. Roa-Morales, E. Peralta-Reyes, E. M. Campo and R. Natividad, Comparison of AOPs Efficiencies on phenolic compounds degradation, J Chem (2016), Article ID 4108587, 8 pages
[14] M. Hoffmann, S. Martin, W. Choi, D. Bahnemann, Environmental applications of semiconductor photo catalysis, Chem Rev,95 (1995) 69-96.
[15] G. Puma, A. Bono, D. Krishnaiah, J. Collin, Preparation of titanium dioxide photo catalyst loaded onto activated carbon support using chemical vapor deposition a review paper. J. Hazard Mater, 157 (2008) 209-219.

IOP Conf. Series: Earth and Environmental Science **78** (2017) 012008 doi:10.1088/1755-1315/78/1/012008

Lead Determination and Heterogeneity Analysis in Soil from a Former Firing Range

Ricardo Urrutia-Goyes[1,*], Ariadne Argyraki[2] and Nancy Ornelas-Soto[3]

1 Universidad de las Fuerzas Armadas ESPE, Sangolqui, Ecuador
2 National and Kapodistrian University of Athens, Athens, Greece
3 Tecnologico de Monterrey, Monterrey, Mexico
*Corresponding author e-mail: erurrutia@espe.edu.ec

Abstract. Public places can have an unknown past of pollutants deposition. The exposition to such contaminants can create environmental and health issues. The characterization of a former firing range in Athens, Greece will allow its monitoring and encourage its remediation. This study is focused on Pb contamination in the site due to its presence in ammunition. A dense sampling design with 91 location (10 m apart) was used to determine the spatial distribution of the element in the surface soil of the study area. Duplicates samples were also collected one meter apart from 8 random locations to estimate the heterogeneity of the site. Elemental concentrations were measured using a portable XRF device after simple sample homogenization in the field. Robust Analysis of Variance showed that the contributions to the total variance were 11% from sampling, 1% analytical, and 88% geochemical; reflecting the suitability of the technique. Moreover, the extended random uncertainty relative to the mean concentration was 91.5%; confirming the high heterogeneity of the site. Statistical analysis defined a very high contamination in the area yielding to suggest the need for more in-depth analysis of other contaminants and possible health risks.

1. Introduction

Bullet cores typically used in firing ranges are made mainly of Pb, which is a known pollutant with many effects on human health. Firing ranges, on the other hand, have shown high concentrations of different trace elements across the globe [1]–[3]. A threat to human health might be created when such firing ranges become restored and available for public use.

Characterizing a site allows to determine the extent and distribution of a suspected contamination in order to define its features e.g. concentration values, spatial distribution, heterogeneity. A proper characterization requires a sampling design that covers the study area as much as possible [4]. Likewise, the spatial heterogeneity of a contaminant in the soil allows to suggest its natural or anthropogenic nature through analyzing the uncertainty of measurements i.e. elements with low variability tend to have natural background concentrations. This can be achieved by estimating the uncertainty empirically from sampling using randomized replicated experiments [5].

A recent technique used for measuring metals in soils and sediments is portable X-ray fluorescence (XRF). This technique has the advantages of scrutinizing the measurements immediately, good detection limits, lower costs of analysis, and keeping an intact sample [6]. Approaches using portable XRF have been applied to soils, sediments, pollution variability, and elemental contamination through cities, among others [7]–[10]. This way, characterizing Pb using portable XRF appears as feasible and necessary.

In this paper, a case study is presented on a former firing range in Athens, Greece. The main focus is the analysis of Pb due to its presence in ammunition. The main objective of this work is to present the spatial distribution of the heavy metal and discuss its heterogeneity characteristics in order to propose its nature. To reach this aim, data from site investigation at an urban park in Athens is used. This type of preliminary site studies are of great help to perform subsequent environmental site assessments.

2. Methodology

2.1. Site Description

The study area is a public place located in the Skopeftirio Park of the Municipality of Kesariani, in Athens, Greece (see Figure 1). The area of the park is ~0.7 km^2, which is mainly covered by coniferous trees and grass. The history of the park began in the 1940's when it was used an execution place and years later when it was used for military purposes. During the 1980's, the same area was used as a shooting range for recreation. Nowadays, it is a historical monument of the city and many people gather around the area for leisure. Along the years, some areas have been modified and others, such as the one here studied, remain intact.

2.2. Instrumentation

Experiments were performed using reagent-grade chemicals and deionized water. Certified reference materials (CRMs) were obtained from NIST® and AccuStandards®. Blank samples consisted of clean silica sand. Portable XRF analysis was carried out using an Olympus Delta Premium 6000 device.

2.3. Sampling

The study area consists of ~0.9 ha. International regulations suggest that such extension can be characterized using a minimum of 10 samples [11]–[14]. However, international guidelines include the preparation of the sample and the use of an analytical technique to define the concentrations of the trace elements. This way, the present study follows a sampling design through a regular grid with 10 m square size, reaching 91 sampling points (see Figure 1).

Figure 1. Locations of 91 sampling points for the study area. Sampling locations in red, additional duplicates in yellow. Referenced on the field with a GPS and long measuring tape. Maps data ©2016 Google.

2.4. Analysis by Portable XRF

Reading time for analysis was 90 s to quantify the target element Pb [15]. Calibration of the portable XRF device was held using CRMs: NIST® 2710, NIST® 2711a, AccuStandard® CRM025-050, and

AccuStandard® CRM023-050. CRMs and a blank sample were read repeatedly before and after the analysis of the soil samples every day. Measurements were taken during the first trimester of 2016.
Soil samples were collected with a spatula to dig through superficial soil. Any vegetation, gravel or debris was removed. An amount of ~400 g was placed in a polypropylene bag and labeled according to the sampling design. Each sample was homogenized by stirring and rotating the sample. Finally, a flat layer was formed and the analyzer was placed on top of the bag. All the tools were cleaned with distilled water in between analyzes to avoid contamination.

2.5. Data Analysis
Data obtained was summarized using mean values, medians, and standard deviations. Calculations were performed using Minitab®. Precision for the portable XRF analysis was estimated by measuring CRMs and calculating the relative standard deviation (RSD). Accuracy of the portable XRF analysis was done by reading CRMs repeatedly [15]. Moreover, in order to perform uncertainty analysis, eight duplicates were collected one meter away from random sampling locations (see Figure 2). Robust ANOVA was applied to the duplicated samples and thus the sampling precision, analytical precision, and geochemical precision were estimated using software package ROBAN from the (UK) Royal Society of Chemistry [16]. Finally, the extended random uncertainty relative to the mean concentration was calculated for Pb using equation (1) for 95% confidence, where s_{meas} is the measurement variance and x is the mean.

$$U_r\% = 200\ s_{meas} / x \quad (1)$$

3. Results and Discussion
Measurements of Pb concentrations in soil samples with a portable XRF device yielded the following results: RSD values ranged from 1 to 2% and mean recoveries were ~110%. Quality control procedures were followed in order to perform the statistical interpretation of the data. The frequency distribution for the concentrations followed a log-normal distribution. Pb values are positively skewed and are generally very high with a mean of more than 2,000 ppm, as can be seen in Table 1. A high variability can be readily seen as the standard deviation and coefficient of variation are also very high.

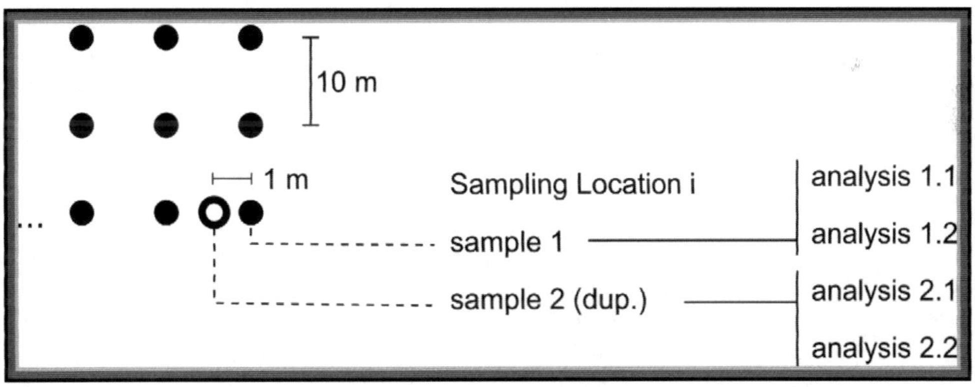

Figure 2. Scheme of duplicate sampling for uncertainty analysis. Modified from Argyraki and Petrakaki [5].

Table 1. Descriptive statistics for Pb based on 91 samples from the study area.

Statistic	Concentration (ppm)
Mean	2,021
Median	649
Standard Deviation	3,606
Skewness	3.725
Minimum	12.5

Maximum	24,824
CV	1.78
SE	378

The nested and duplicate sampling design (see Figure 2) was used in order to estimate the random measurement error for Pb. Sampling and analytical quality control was also applied. Table 2 shows the concentrations used in Robust Analysis of Variance applied to the Pb sampling and analytical duplicate samples in order to estimate the percentage of the total variance that can be attributed to the geochemical variance, sampling variance, and analytical variance. Such proportions were calculated and displayed in pie charts (see Figure 3). As shown in Table 2, the analytical variance (<1% of total variance) is lower than the maximum reference value of 4%. Similarly, the sampling variance (~11% of total variance) is lower than the reference value of 20% [17]. Both reference values work as a guide for an acceptable spatial interpretation of the element distribution. Finally, it can be noted that the high geochemical variance (~88%) is the main component of the total variance, which in turn will lead to a high degree of uncertainty. In fact, the extended random uncertainty relative to the mean concentration was calculated as 91.5%. This suggests that the nature of the trace element is indeed anthropogenic

It has been thought that Pb is only present in small spots across the study area [5],[18], however, it has been demonstrated that the contamination of the soil is extremely high and widespread. The dispersion of the element in the media, mainly due to erosion, is present in spots bigger than tens of meters wide. A very dense sampling design has been very useful to determine the distribution of a contaminant with such a high heterogeneity. It can also be stated that a very low level of analytical variance indicates that chemical analysis is not a source of error during the process and hence that the portable XRF device appears to be suitable for geochemical analysis in this case.

Table 2. Concentrations (ppm) of Pb according to the duplicate scheme from Figure 2 and results of Robust Analysis of Variance

Sample	Analysis 1.1	Analysis 1.2	Analysis 2.1	Analysis 2.2
C2	23,049	23,034	25,452	25,467
D4	2,445	2,433	1,067	1,069
D7	2,248	2,219	2,316	2,312
F5	8,020	8,097	6,976	6,963
F7	174	169	59	57
H3	1,629	1,641	176	180
H8	21	24.1	200	197
J9	24.7	24.9	301	292
K11	43.9	45.0	25.2	22.2
Robust ANOVA	Between target	Sampling	Analysis	Measurement
Std. deviation	2085.8	747.3	7.5	747.3
% of total var.	88.6	11.3	0.0	11.3

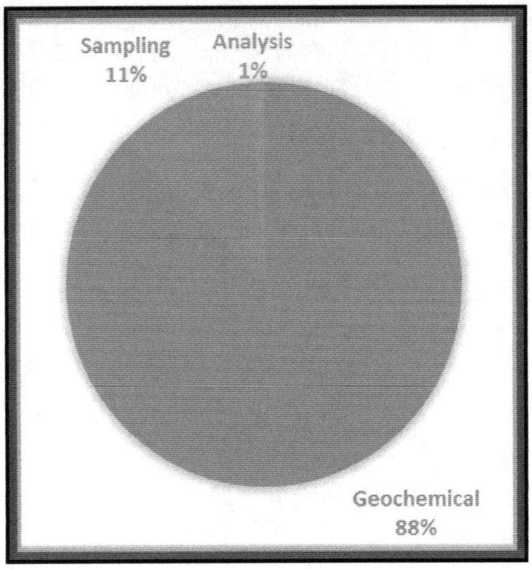

Figure 3. Sampling, analysis, and geochemical proportions of total variance for Pb using Robust ANOVA showing the relative importance of measurement errors.

The concentrations obtained and calculations performed allow to witness a realistic spatial interpretation of Pb across the study area. However, the high level and wide range of concentrations measured suggest that verifying some measurements can be useful. In this instance, the preparation of the sample and subsequent analysis using a traditional analytical technique will allow to compare and contrast the accuracy of the portable XRF technique. Likewise, a human health risk analysis will allow to determine the degree of risk to which the population is exposed by using the public area here studied.

4. Conclusion

A public space from Athens, Greece has been characterized for Pb contamination using a portable XRF device. A 10 m square grid used in this survey proved to be suitable to characterize Pb contamination using a 20% criterion. Robust Analysis of Variance performed over a nested design of sampling and analytical duplicates has allowed the determination of the uncertainty of the contaminant in the field. High Pb concentrations with high spatial heterogeneity have been reported. Results regarding Pb contamination will hopefully yield concerns for further studies.

5. Acknowledgments

The authors acknowledge the collaboration in the project of Dr. Martin Bremer-Bremer and Mucio Rodriguez from Tecnologico de Monterrey; the Greek Institute of Geology and Mineral Exploration; and the funding of Consejo Nacional de Ciencia y Tecnología (CONACYT), Mexico (scholarship #387660).

6. References

[1] Okkenhaug G, Grasshorn Gebhardt K-A, Amstaetter K, Lassen Bue H, Herzel H, Mariussen E, Rossebø Almås Å, Cornelissen G, Breedveld G D, Rasmussen G and Mulder J 2016 Antimony (Sb) and lead (Pb) in contaminated shooting range soils: Sb and Pb mobility and immobilization by iron based sorbents, a field study. *J. Hazard. Mater.* 307 336–43

[2] Ackermann S, Gieré R, Newville M and Majzlan J 2009 Antimony sinks in the weathering crust of bullets from Swiss shooting ranges *Sci. Total Environ.*

[3] OSHA 2014 Safety and Health Topics Occup. adn Heal. Top.

[4] Sorvari J 2011 *Encyclopedia of Environmental Health* (Elsevier)

[5] Argyraki A and Petrakaki N 2010 Heterogeneity in heavy metal concentrations in the soil of a firing range area at Kesariani, Athens, Greece Proceedings of the 12th International Congress of the Geological Society of Greece, Planet Earth: Geological Processes & Sustainable Development vol XLIII (Patras: Bulletin of the Geological Society of Greece)

[6] Ramsey M H and Boon K a. 2012 Can in situ geochemical measurements be more fit-for-purpose than those made ex situ? *Appl. Geochemistry* 27 969–76
[7] Peinado F M, Ruano S M, González M G B and Molina C E 2010 A rapid field procedure for screening trace elements in polluted soil using portable X-ray fluorescence (PXRF) *Geoderma* 159
[8] Mejía-Piña K G, Huerta-Diaz M A and González-Yajimovich O 2016 Calibration of handheld X-ray fluorescence (XRF) equipment for optimum determination of elemental concentrations in sediment samples *Talanta* 161
[9] Weindorf D C, Paulette L and Man T 2013 In-situ assessment of metal contamination via portable X-ray fluorescence spectroscopy: Zlatna, Romania. *Environ. Pollut.* 182 92–100
[10] Clark J J and Knudsen A C 2013 Extent, Characterization, and Sources of Soil Lead Contamination in Small-Urban Residential Neighborhoods *J. Environ. Qual.* 42 1498–506
[11] CCME 1993 Guidance Manual on Sampling, Analysis, and Data Managemenent for Contaminated Sites (Winnipeg: CCME)
[12] NSW EPA 1995 Contaminated Sites: Sampling Design Guidelines (Sydney)
[13] SEMARNAT 2006 NMX-AA-132-SCFI-2006 (Mexico)
[14] EPA Victoria 2009 Industrial waste resource guidelines: Soil sampling (Melbourne)
[15] USEPA 2007 Method 6200 (United States)
[16] Ramsey M H and Ellison S L R 2007 Eurachem/EUROLAB/CITAC/Nordtest/AMC Guide: *Measurement uncertainty arising from sampling: A guide to methods and approaches*
[17] Ramsey M H 1993 Sampling and analytical quality control (SAX) for improved error estimation in the measurement of heavy metals in the environment, using robust analysis of variance Appl. Geochemistry 2 149–53
[18] Petrakaki N 2009 Lead distribution in firing ranges: the case of Skopeftirio Park in Athens (University of Athens)

Gas Mitigation in Paper Production

Santos AS[1]and BittencourtC[2]

1 Graduated in the Environmental and Sanitary Engineering Area, by the Universitary Center of the Metropolitan University Faculties - FMU. 06/2017. Technique WorkSafetyby SENAC - SP, 2011.

2 PhD student at the Energy and Nuclear Research Institute of the University of São Paulo (IPEN / USP). Specialist in Environmental Pollution Control Engineering at the Faculty of Public Health (FSP / USP), postgraduate degree in Sewage Treatment by TUHH Hamburg / Germany, graduated in Chemical Engineering - Faculties Oswaldo Cruz (1992). She is currently a professor of Graduate and Undergraduate courses at Universitary Center of the Metropolitan University Faculties – FMUand an engineer at the Basic Sanitation Company of the State of São Paulo, working on the regulation of the sanitation and service sector of the secretariat. Author of the book Water Treatment and Effluents: Fundamentals of environmental sanitation and water resource management and the book Regulatory Agencies in Brazil.
E-mail: andreass2002br@gmail.com;g.enge.cb@gmail.com

Abstract. The Brazilian paper industry has competitive advantages offered by the favorable climate, which favors an increase in the yield of forest restoration, and consequently, in the productive process. On the other hand, following the greenhouse gases (GHG), we can see our constantly changing sun, causing the solar storms, allowing their prevention or mitigating measures. The objective of this work is to contribute to the construction of the understanding necessary for the reduction of GHG emission from a preliminary analysis of the pulp and paper sector. As a secondary objective, the text preliminarily analyzes a company's behavior against the backdrop of the Paris Accord, which strengthens the global response to the threat of climate change and strengthens the capacity of countries to deal with the impacts of such changes. The identification of best practices in the pulp and paper industry is understood, focusing on environmental sustainability, such as the adoption of reforestation, obtaining significant results. In the case of the paper industry, the management of public forests for sustainable production, within the structure of the Ministry of the Environment, establishes the promotion of public awareness about the importance of conservation, recovery and sustainable management of forest resources.

1. Introduction

Paper is considered the main support for the diffusion of writing, information and all human knowledge. Before the discovery of paper, man used the most different materials for the record of his existence, such as leaves, bark, leather, cloths, stones, clay and metals. The paper appeared in China at the beginning of the second century, invented by T'SaiLum, an officer of the Court who would have fabricated the paper from the cortex of plants, old textiles and fragments of fishing nets. [1].
The company Fibria was chosen for the job, as a potentially polluting company, we will analyze the Environmental Performance Index (IDA), which is a tool that evaluates the quality of the processes in

the industry through indicators such as pollution prevention and control and environmental management, verifying that it is actually meeting the Forest Code with the restoration in the field.
This work has as objective:
1. Contribute to the construction of the understanding necessary for the reduction of greenhouse gas emissions (GHG) from the pulp and paper sector.
2. Preliminary analysis of the behavior of a company against the scenario indicated by the Paris Agreement.
Part of the solar energy that comes to the planet is reflected directly back into space, reaching the top of the Earth's atmosphere, and part is absorbed by the oceans and the Earth's surface, promoting its heating. A portion of this heat is radiated back into space, but is blocked by the presence of greenhouse gases that, despite passing energy from the Sun (emitted at smaller wavelengths), are opaque to terrestrial radiation emitted in larger Wavelengths. This difference in wavelengths is due to differences in the temperatures of the Sun and the Earth's surface.
The presence of these gases in the atmosphere makes it habitable, because if they did not exist naturally, the average temperature of the planet would be very low, on the order of minus 18°C. The energy exchange between the surface and the atmosphere maintains the current conditions, which provide an average global temperature near the surface of 14ºC. [2].
Our ever-changing sun releases solar material into space. The grandest events are massive clouds emerging from the sun, called coronal mass ejections (CME). These solar storms come first with some kind of warning - the flash of a bright flash, a blast of heat or a flood of solar energy particles. But another kind of storm has puzzled scientists for their lack of typical warning signs: They seem to come out of nowhere, and scientists call it a stealthy CME. [3].
To discover the origins of stealthy CME, scientists have developed a model of the sun's magnetic fields, simulating its strength and motion in the sun's atmosphere. Central to the model was the differential rotation of the sun, which means different points about the sun rotating at different speeds. Unlike the Earth, which rotates like a solid body, the sun rotates faster at the equator than at its poles. [3].
Earth's magnetosphere is created by our magnetic field and protects us from most of the particles the sun emits, these solar eruptions are difficult to predict and may interact with space weather effects.
Space time studies provide us with the ability to predict events and conditions in the Sun and near Earth, preceding precision to allow for prevention or mitigating measures to be taken.
The results of global assessments of the Third Assessment Report (TAR) of the Intergovernmental Panel on Climate Change (IPCC 2001) Working Group 2 (WG2) on impacts, adaptation and vulnerability to climate change at the regional level can be summarized as follows:) Recent climate changes, especially rising temperatures, are already affecting physical systems (climate, water resources) and biological systems (ecosystems, human health, cities, industries); B) Preliminary indicators exist that some human systems have already been affected by drought or floods; C) Natural systems are vulnerable to climate change, and some will be irreversibly damaged; D) Those with less resources and less able to adapt are the most vulnerable. [4].
Although Brazil's contribution to the global concentration of greenhouse gases is lower than that of the industrialized countries, and contribution due to fires (smoke and aerosols) are quite high. Projections of climate models allow the generation of climate scenarios in the future, but it does not yet distinguish or separate the effects of natural climate and man-induced variability. [4].
The Brazilian Association of Trees (IBÁ) identified growth in its volume of exports of the three segments analyzed, ie pulp, wood panels and paper, compared to the same period of the year 2015. IBÁ identified a positive result, Which is reflected in the trade balance of the forestry sector, which reached US $ 6 billion, representing growth of 2.3%. [5].
In the same period, Brazilian pulp production reached 17.1 million tons and paper pulp 9.5 million tons. The main destinations for Brazilian exports of paper and wood panels are Latin American countries. In the case of Brazilian pulp, the largest importing country in 2016 was China. [5]
Brazilian pulp production rose 4.8% in January 2017 over the same period in 2016 to 1,665 million tons, with exports of the raw material rising 47.4% in the same comparison. Sales of wood panels in the country, an important input for civil construction, rose by 7.6% in January from 1 year earlier, to

508,000 m³. Paper production in January fell 0.5% from a year earlier to 862,000 tonnes, with domestic sales falling 4.9% in the same period to 424,000 tonnes. [6].
Based on these data, we can observe the growth of the Gross Domestic Product - Brazilian GDP, together with the GHG, in evidence the carbon due to the materials used, the association of the industries has great potential to contribute to the positive interferences to the climatic changes, seeking to exploit this market for a low carbon economy.

2. Comparison between Brazilian Production and that of Other Countries

Currently, the most used vegetable raw material in papermaking is wood, although others may also be used. These raw materials are now processed chemically or mechanically, or by a combination of the two modes, generating as a product what is called cellulosic pulp, which can be further bleached if a white pulp is desired. The cellulosic pulp is composed of the cellulosic fibers improving the bond between them, after undergoing a thinner purification to separate the existing sands in the pulp. Finally, the pulp is bleached with chlorine or peroxide compound following to the papermaking machines. [7].
China and the United States, despite being the largest producers, do not produce enough to meet domestic demand. Brazil is in 4th place among the world's largest pulp producers, is expected to climb to third place, surpassing Canada in the next six years, according to the plan to expand production in the country. [8]. (see Table 1)

Table 1. The 10 largest pulp producers in the world.

Position	Country	production
1°	United States	18. 308 tonnes
2°	China	19.542 tonnes
3°	Canada	18. 308 tonnes
4°	Brazil	13.922 tonnes
5°	Sweden	11.859 tonnes
6°	Finland	10.363 tonnes
7°	Japan	9.020 tonnes
8°	Russia	7.453 tonnes
9°	Indonesia	6.805 tonnes
10°	Chile	4.876 tonnes

Source: (Lairtes Chaves, 2012)

Table 2. In 2012, the 10 largest countries that emit greenhouse gases account for more than two-thirds of total global emissions. [9]. (see Table 2)

Position	Country	Production	%
1°	China	10.684.29 MT CO2e	22.44%
2°	United States	5.822.87 MT CO2e	12.23%
3°	European Union	4.122.64 MT CO2e	8.66%
4°	India	2.887.08 MT CO2e	6.06%
5°	Russia	2.254.47 MT CO2e	4.73%
6°	Indonesia	1.981 MT CO2e	4.16%
7°	Brazil	1.823.15 MT CO2e	3.83%

8º	Japan	1.207.30 MT CO2e	2.53%
9º	Canada	856.28 MT CO2e	1.79%
10º	Mexico	748.91 MT CO2e	1.57%

Source: (Johannes Friedrich, Mengpin Ge and Thomas Damassa, 2015).

3. From the Paper Industry, from the Brazilian Industrial Park

Brazil is the first producer of e pinus eucalyptus pulp, where sanitary products are manufactured: toilet paper, disposable diapers, absorbents and napkins, among other related items, called short fiber pulp. The favorable climate had a great role in the development of the national tree engineering, and with that, the high forest yield of Brazil added to the exchange advantage. [10].

The 2010 greenhouse gas inventory of Brazil shows that the total amount of carbon dioxide (CO2) launched by the country is reduced by 50% between 2005 (previous inventory) and 2010, from 2.73 billion tons of CO2 to 1.27 billion tonnes. By the calculation, in 2005, when deforestation in Brazil was at its peak, 2.73 billion tons of CO2 were emitted, in the second inventory 2.1 billion tons of CO2 were accounted for. In 2005 it was the base year used by the government to propose greenhouse gas emission reduction targets submitted to the Paris Agreement. [11].

Paper and pulp companies should be oriented towards a production with sustainability in the economic, environmental and social aspects that need to interact as a whole, to guarantee the permanence of the business, without compromising the environment. [12].

The Agreement recognizes that climate change is a common concern of mankind and requires broad cooperation from all countries to accelerate the reduction of global GHG emissions. To contribute to the mitigation of GHG emissions and sustainable development, it shall establish under the authority, guidance and meeting of the Conference of the Parties to the Paris Agreement its supervision.

The Paris Agreement was approved by the 195 UNFCCC Party to reduce greenhouse gas (GHG) emissions in the context of sustainable development. To strengthen the implementation of the Convention, including its objective, which aims to strengthen the global response to the threat of climate change. The commitment is to maintain the overall average temperature increase below 2°C above pre-industrial levels and to make efforts to limit this temperature increase to 1,5°C in relation to pre-industrial levels, That this would significantly reduce the risks and impacts of climate change. [13].

After approval by the National Congress, Brazil concluded, on September 12, 2016, the process of ratification of the Paris Agreement. On September 21, the instrument was delivered to the United Nations. As a result, Brazilian targets were no longer intended and became official commitments. Now, therefore, the acronym has lost the letter "i" (from English, intended) and is now called the NDC only. [13].

4. Company Studied Fibria

The planted trees sector will play a key role in the goals of the Paris Agreement, the Brazilian partnership becomes part of the agreement, which sets the commitment for low carbon emissions, Brazil becomes the third major issuer to confirm the participation of the treaty.

The Fibria company is part of this sector together with IBÁ, the solution passes through planted forests, with a 100% renewable base, it is estimated that the mitigation potential of the effects of climate change is directly proportional to the capacity of creation and use of mechanisms in the Carbon market, together with integrated and coordinated public policies.

With the Forest Code, the company becomes a producer of renewable and sustainable forests, developed for the production of pulp and paper. The results are expected as a result of the possibility of evaluating the identification and management of its operation risks, its aspects and impacts to the environment Environment, society, health and safety of professionals and the quality of its products and services. The company has developed a prevention strategy aligned with the recommendations of the Kyoto Protocol, the UN Conference of the Parties (COPs) and the National Policy on Climate Change, among other forums.

The IDSA is the tool that the company uses to identify good practices and fragilities related to the socio-environmental performance of the Forest area, promoting maintenance and joint improvements through the interaction of the environment with the operational areas.
Seeking to anticipate and address the legitimate concerns of stakeholders before requiring costly mitigation measures. Identifying environmental or social improvements in forest management, pulp production and logistics that also provide financial gains. Anticipating as much as possible potential issues and risks, gaining time for planning and financing (important in the pulp industry) when changes are needed. [14].
By making use of the treaty signed with the Paris Agreement with Brazil, the company undertakes to:
1. Publish annually the inventory of emissions of greenhouse gases (GHG) of companies, as well as actions for mitigation of emissions and adaptation to climate change.
2. Include as strategic orientation in the investment decision-making process the choice of options that promote the reduction of GHG emissions in processes, products and services.
3. Seek the continuous reduction of specific GHG emissions and the net balance of CO2 emissions of companies through actions to directly reduce emissions in production processes, investments in carbon capture and sequestration and / or support for reduction actions Emissions from deforestation and degradation. [15].

5. Gas Control Legislation

With a production capacity of 5.3 million tons per year, the company has industrial units located in Aracruz (ES), Jacareí (SP) and Três Lagoas (MS), in addition to Eunápolis (BA), where Veracel maintains joint- Operation with Stora Enso. The company has 969 thousand hectares of forests, 568 thousand ha of planted forests, 338 thousand ha of preservation and environmental conservation areas and 63 thousand ha for other uses. The pulp produced by Fibria is exported to more than 40 countries. [16].
Brazilian environmental legislation is quite sophisticated and old, we can highlight Law 6.938 of August 31, 1981, which establishes the National Environmental Policy (PNMA) and Decree No. 99,274 of June 6, 1990, which regulates it. Among the guidelines established by it are: The establishment of environmental quality standards; The assessment of environmental impacts; and The licensing of polluting activities.
Established by Law No. 12,187 of December 29, 2009, the National Policy on Climate Change (PNMC) establishes its principles, objectives, guidelines and instruments.
State Law No. 13,798 of November 9, 2009, the State Policy on Climate Change - PEMC, regulated by State Decree No. 55,947, of June 24, 2010, is in line with the UN Climate Convention and the National Policy on Climate Change. Climate Change.
The National Plan for Adaptation to Climate Change, as established by Ordinance No. 150 of May 10, 2016, is coordinated by the Technical Adaptation Group, created with the function of providing technical and political guidance for actions under this Plan, Monitor it, evaluate it and review it, as well as establish the detailed routines and operational mechanisms for its management. [17].
Law No. 11,284 of March 2, 2006, the Public Forest Management Law, which provides for the management of public forests for sustainable production, establishes the Brazilian Forest Service (SFB), within the structure of the Ministry of the Environment, and creates the National Forest Development Fund (FNDF). Article 2 establishes the principles of public forest management, which may mention the protection of ecosystems, soil, water, biodiversity and associated cultural values; Promotion of local processing and encouragement of increased value added to forest products and services as well as industrial diversification, technological development, use and capacity building of local entrepreneurs and regional labor; The promotion of awareness and the promotion of public awareness about the importance of conservation, recovery and sustainable management of forest resources.
The Forest Code, reformulated in October 2012, establishes general norms on vegetation protection, Permanent Preservation areas and Legal Reserve areas; Logging, supply of forest raw materials, control of the origin of forest products and control and prevention of forest fires, and provides economic and financial instruments to achieve its objectives.

Subak [18], evaluated the pulp and paper industry worldwide, considering the pulp and paper production process from the removal of wood from the forest to the final disposal in landfills. Fiber supply depends on trees, papermaking requires fuel inputs and paper waste disposal can contribute to greenhouse gas (GHG) and methane (CH_4) emissions, a GHG is twenty times more potent than Or CO_2.
The authors conclude that overall the sector emits around 460 million tonnes of CO_2 equivalent, more than is fixed by the growth of forests, it is estimated that the contribution to global warming of landfill paper is similar to that of [18], as the most significant in the sector's emissions balance, which has achieved important advances in the mitigation of emissions from energy use and biomass disposal, [18] has changed considerably and the results need to be updated to clearly assess the positioning of the Brazilian pulp and paper sector in relation to the current global scenario, but the comparison with the reality of a decade ago is Useful in assessing the correctness and opportunities for improvement that still exist in the sector in relation to GHG mitigation.
Regarding the forest base, the practices adopted with success were the investments in the development of technologies that allowed to reach records of productivity of wood with the use of the natural potentialities of the country. The role of the most recognized forest certification, the Forest Stewardship Council (FSC) and the National Forest Certification Program (Cerflor), endorsed by the International Program for the Endorsement of Forest Certification Systems (PEFC), should be highlighted here. Of the total hectares of trees planted in Brazil, 58% are certified by these entities. [19].
In face of the legislation and norms evaluated, the concern with the environment is identified not only in the present but also in the future. Laws were created to improve resources, certifying companies with sustainable policies, with the Paris Agreement, the partnership with the countries with the same objective was evident, with no measures to be put into practice.
With respect to the generation of GEE, two models prevail that describes the subject, the first one: says to be good for providing conditions for the development of life, through nitrous oxide giving life to the planet. The second against says the first, informing about its impacts on climate change, in the physical, biological scopes. Each author has a different view of the other not agreeing with the presented parameters, the important thing is to be aware of the aspects and impacts, so that they are controlled.
Adopting the perspectives regarding the sustainability of the processes, the economic, environmental and social fields are naturally adopted, interacting so as not to compromise the environment, the paper and pulp companies can guarantee the continuity of the process.
In the observed analysis of the company, it is being structured so that the program of reduction of the impacts of climate change, through actions that cover the entire product life cycle, expects to sensitize its partners to the theme and achieve positive results for the environment.
With its increase in forest areas, by doubling the carbon absorption of the atmosphere, restoring degraded areas with native species, collaborating with social and environmental responsibilities, producing its raw material, performing integrated management of standards, managing its risks Operation. Seeking to anticipate and address the legitimate concerns of stakeholders before requiring costly mitigation measures.

6. Conclusion

Based on information provided by the Brazilian Tree Industry (IBÁ), the pulp, paper and paper artifact industries have different configurations and evolutions in the Brazilian economy.
Considering the opportunities for expansion of planted and native forests, the carbon market becomes capable of economically enhancing climatic benefits, combating climate change becomes possible, together with public policies, that will put into practice the fundamentals and principles To ease, as we have followed the evolution of the effect of GHG.
We can see our constantly changing sun causing solar storms, with monitoring constantly being made possible its prevention or mitigating measures.
According to the evaluations carried out, Fibria developed a prevention strategy in line with the recommendations of the Kyoto Protocol, the UN Conference of the Parties (COPs) and the National

Policy on Climate Change, among others, using the IDA that identifies good Practices and weaknesses related to the socio-environmental performance of the Forest area and promoting maintenance and improvements.
The company undertakes to publish its annual inventory of GHG emissions, mitigating its internal processes such as: conservation importance, recovery and sustainable management of forest resources, protection of vegetation, fauna, flora, Permanent Preservation areas and Legal Reserve areas, control of the forest exploitation taking advantage of new business opportunities and increasing its competitiveness.
Including strategic guidelines in the GHG processes and services, maintain the search for continuous reduction of specific GHG emissions and the net balance of CO2 emissions of companies through actions of direct reduction of emissions in production processes and services.
To reduce GHG emissions, the GHG is committed to meeting the growing global demand for forest products, the Paris Agreement, contributing to global efforts by monitoring, evaluating and reviewing for integrated management.

7. References

[1] Arroja L, Dias A C and Chapel I (2006) The role of Eucalyptus globulus Forest and products in carbon sequestration.Climatic Change

[2] Ministry of Environment. Greenhouse Effect and Global Warming. Retrieved from http://www.mma.gov.br/clima/energia/item/195-efeito-estufa-e-quequecimento-global Accessed on: 06 May 2017

[3] NASA Space Weather: The Space Time model simulates solar storms from nowhere. Retrieved from https://www.nasa.gov/feature/goddard/2017/nasas-fermi-sees-gamma-rays-from-hidden-solar-flaresAccess on: 10 May 2017

[4] Ministry of the Environment. Report 1 Characterization of the climate in the 20th Century and Climatic Scenarios in Brazil and South America for the 21st Century derived from the Global Climate Models of the IPCC. Retrieved from **http://mudancasclimaticas.cptec.inpe.br/~rmclima/pdfs/prod_probio**/Relatorio_1.pdf Accessed on: 10 May 2017

[5] CEPEA Center for Advanced Studies in Applied Economics. Brazil exports US $ 6 billion in pulp, paper and wood panels in the first eleven months of 2016. Retrieved from http://www.cepea.esalq.usp.br/br/categoria/acessar/informativo-florestal-1.aspx. Accessed on: 28 feb. 2017

[6] Exame.com.Production of Pulp in Brazil in January has an annual increase of 4.8%.Retrieved from http://exame.abril.com.br/economia/producao-de-celulose-do-brasil-em-janeiro -to-high-annual-of-48. Accessed on: 28 Feb. 2017

[7] Santos G P S, Alves D F, Paiva L S and Nunes R V A (2010) Paper and paperboard chain and recycling: a comparative analysis in the packaging industry. In: National Meeting of Production Engineering, XXX, 2010, São Carlos, São Paulo

[8] Forest Panel. The 12 countries that most produce pulp in the world. Retrieved from http://www.painelflorestal.com.br/noticias/celulose-e-papel/os-12-maiores-produtores-de-celulose-do-mundo Accessed on: 01 May 2017

[9] CAIT Climate Data Explorer. Infographic: How are the emissions of your country? Retrieved from http://www.wri.org/blog/2015/06/infographic-what-do-your-countrys-emissions-look Accessed on: 01 May2017

[10] The globe. China's demand for hygiene products boosts paper exports. Retrieved from http: //oglobo.globo.com/economy/business/cambio-demanda-de-china-por-producto-de-higiene-impulsionam-exportacao-de-papel-18630424 Accessed on: Feb 28th. 2017

[11] Exame.com. Emissions of polluting gases in Brazil fell 50% in 5 years. Retrieved from http://exame.abril.com.br/brasil/emissoes-do-brasil-caem-50-nos-ultimos-5-anos/. Accessed on: 28 feb. 2017

[12] Rosa B N and Moraes G G (2005) The importance of recycling paper in improving the quality of the environment. In: National Meeting of Production Engineering - ENEGEP, Anais, 25., 2005. Porto Alegre, RS

[13] Ministry of the Environment. Foundations for the Elaboration of the Desired Nationally Determined Contribution (iNDC) of Brazil in the context of the Paris Agreement. Retrieved from http://www.mma.gov.br/clima/convencao-das-nacoes-unidas/acordo-de- Paris / itemlist / category / 138-climate-convention-on-climate-change. Accessed on: 28 feb. 2017

[14] Fibria Institutional Sustainability. Retrieved from **http://www.fibria.com.br/web/pt/institucional/sustentabilidade.htm**. Accessed on: 28 feb. 2017

[15] Fibria Environment Climate Change. Retrieved from http://www.fibria.com.br/ambiente/mudancas-climaticas/.Accessed on: 28 feb. 2017

[16] Fibria Media Releases. January 31, 2017 - Fibria ends 2016 with sales growth, strong net profit and robust cash position. Retrieved from **http://www.fibria.com.br/midia/releases/fibria-encerra-2016-com-crescimento-de-vendas-forte-lucro-liquido-e-robusta-posicao-de-caixa/** Accessed on: 28 feb. 2017

[17] Ministry of the Environment. Retrieved from http://www.mma.gov.br/clima/adaptacao/plano-nacional-de-adaptacao Accessed on: 28 Feb.2017

[18] Subak S and Craighill A (1999) The contribution of the paper cycle to global warming.Mitigation and Adaption Strategies for.Global Change

[19] IBÁ Brazilian Tree Industry.Ibá certification. Retrieved from http://iba.org/en/arvores-plants/certificacao. Accessed on: 28 feb. 2017

Insecticide Usage and Chemical Contamination Assessment in Asiatic Pennywort

S Bumroongsook
Department of Plant Production Technology, Faculty of Agricultural Technology
King Mongkut's Institute of Technology Ladkrabang, Bangkok 10520, Thailand
Email: suvarin.bu@kmitl.ac.th

Abstract. The insecticide usage in commercially grown asiatic pennywort plantations in Nakhonpatum and Nonthaburi province,Thailand was surveyed during January-June, 2016. The results showed that asiatic pennywort cuttworms was leaf destructive and caused the most damge to the production. The growers used organophosphate insecticides to control the caterpillars the most, followed by pyrethoid, abamectin, carbamate and organochlorine, respectively. The chemical contaminants of pennywort from 9 fresh markets in Bangkok was monitored, the result indicated that lead was not detected in the samples. The amount of arsenic was less than 0.075 mg / kg. The insecticide residue measurement of dicofol, chlorpyrifos and methidathion was 0.98, 2.84 and 0.46 mg / kg, respectively.

1. Introduction

Asiatic pennywort is health benefit and has been used for various treatment. The major problem for pennywort production for local consumption and export is damage and destruction caused by cutworms (*Zonoplusia ochreata* Walker). The sizes of these insect populations vary with seasons and cultivated areas. Infestation of common cutworm causes considerable decrease in pennywort production [1].

Insect pest control by insecticide is the most widely used for control method, producing quick and clear results, but chemicals have a huge impact on ecosystem and destroy natural enemies of insect pests. Up to 33% of pennywort from markets was pesticide contaminated and was recorded the most contaminated vegetable among samples[2]. Thai Department of Agriculture banned pennywort production from export due to a residue problem in 2011.

2. Material and Methods

2.1. Questionaire Survey of Insecticide Usage

The questonaire survey was conducted to interview growers of commercialy grown Asiatic pennywort from Nakhonpatom and Nonthaburi province. The growers were randomly selected for the participation at 30 persons from each province to achieve the information on pest problems and insecticide usage.

2.2. Lead and Arsenic Analysis

Asiatic pennywort sample was collected from 9 location of fresh markets in Bangkok. The sample of 3 g of powdered pennywort was digested with nitric acid till the solution was clear and measured. Determination of heavy metals (official method 986.15 for arsenic and official method 999.10 for lead by ICP-MS was based on AOAC[3]. Then, it was compared with standard substances and the arsenic and lead content was calculated.

2.3. Organochlorine and Organophosphate Determination
The sample for this analysis was collected from the same place as in 2.2. The analysis of organochlorine and organophosphate residues were performed followed Steinwandter [4]. Universal 5 minutes online method for extracting and isolating insecticide residues based on the Fresenius Journal of Analytical Chemistry no 1155. The analysis was performed by the GC-MS.

3. Results and Discussion

3.1. Pest Problems and Insecticide Usage
The interview results showed that growers from both Nonthaburi and Nakhonpatum had similar pests and the most destructive one was the pennywort cutworm(Zonoplusia ochreata(Walker) followed by diamondback moth, plant hoppers and thrips(see Table 1). The pennywort cutworm was a leaf feeder and could devour the entire plants. Therefore, the grower preferred chemical control method due to fast and easy. The most preferred insecticide group was organophosphate(43.33% of Nakhonpatom and 33.33% of Nonthaburi)(Table 2).

3.2. Lead and Arsenic Contamination
The content of lead and arsenic was calculated from the standard curve: Y=2.542E+0003*x+1.474E+002(r^2 = 0.9999) and Y=4.440E+0003*x+1.610E+003(r^2 = 0.9999), respectively. No lead detection on pennywort collected from fresh markets in Bangkok. Arsenic was found 20% of samples with less than 0.075 mg / kg(Table 3).

Table 1. Insect pest of Asiatic pennywort

Insect pests	Percentage of growers(n=30)	
	Nakhonpatum	Nonthaburi
pennywort cutworm	43.33	53.33
diamondback moth	26.67	26.67
plant hoppers	16.67	13.33
thrips	13.33	6.67

Table 2. Insecticide usage in Asiatic pennywort plantations

Insecticide group	Percentage of growers(n=30)	
	Nakhonpatum	Nonthaburi
abamectin	26.67	30.00
carbamate	20.00	26.67
organophosphate	43.33	33.33
pyrethoid	3.33	10.00
organochlorine	6.67	0.00

Table 3. Lead and arsenic contamination on asiatic pennywort collected from local markets in Bangkok.

Heavy metal	amount[1](mg/kg)
lead	ND
arsenic	<0.075

[1]ND= not detedcted

3.3. Organochlorine and Organophosphate Contamination

The organochlorine group in pennywort was analysed and the results showed that only dicofol was found 0.98 mg/kg and above the MRL of 0.1 mg/kg. No detection was found on BHC, heptachlor&heptachlor epoxide, aldrin&dieldrin, DDT, chlordane, endosulfan and endrin(Table 4). Detemination of organophosphate group showed that the samples had chlorpyrifos and methidathion 2.84 and 0.64 mg/kg, respectively. The sample contained the chlorpyrifos residue above the MRL of 0.5 mg/kg[5]. No detection was found on DDVP, methamidophos, mevinphos, omethoate, diazinon, dicrotophos, monocrotophos,dimethoate, pirimiphos-methyl, chlorpyrifos, parathion-methyl, pirimiphos-ethyl, malathion, malathion, fenitrothion, parathion ethyl, prothiofos, methidathion, profenofos, triazophos, EPN, phosalone and azinphos-etthy(Table 5).

Table 4. Organochlorine insecticide contamination on asiatic pennywort collected from local markets in Bangkok.

Organochlorine insecticides	amount[1](mg/kg)
BHC	ND
Heptachlor&Heptachlor epoxide	ND
Aldrin&Dieldrin	ND
Dicofol	0.98
DDT	ND
Chlordane	ND
Endosulfan	ND
Endrin	ND

[1]ND = not detected

Table 5. Determination of organophosphate insecticide contamination on

Asiatic pennywort

Organophosphate insecticides	Amount[1](mg/kg)
DDVP	ND
Methamidophos	ND
Mevinphos	ND
Omethoate	ND
Diazinon	ND
Dicrotophos	ND
Monocrotophos	ND
Dimethoate	ND
Pirimiphos-methyl	ND
Chlorpyrifos	2.84
Parathion-methyl	ND
Pirimiphos-ethyl	ND
Malathion	ND
Fenitrothion	ND
Parathion ethyl	ND
Prothiofos	ND
Methidathion	0.46
Profenofos	ND
Triazophos	ND
EPN	ND
Phosalone	ND

Azinphos-etthyl	ND

[1]ND = not detected

4. Conclusions

Growers rely on chemical method to control insect pest due to fast and easy way. They continue the calendar spray and subsequently induce pest resurgence and pennywort product contamination. The residues of dicofol and chlorpyrifos in pennywort were exceed acceptable levels. Moreover, pennywort was not only food but also for pharmaceutical products. Therefore, it should be free of pesticides. Asiatic Organic farming should be implement for sustainable production.

5. References

[1] Ngeonyoo P and S Bumroongsook. 2011. Prof. of the 10th National Horticultural Congress (Bangkok:Thailand)
[2] Thai FDA 2011 The prelimnary analysis of chemical residues and microorganisms Food and Drug Administration, Ministry of Public Health
[3] AOAC. 2005. *Official Method of Analysis*. (Madison: AOAC International)
[4] Steinwandter H 1985 Analytical Methods for Pesticide and Plant Growth Regulators vol XVII, ed J Sherma(New York: Academic Press INC)
[5] National Bureau of Agricultural Commodity and Food Standards 2008 Pesticide Residues: maximum residue limit *Royal Gazette* 125: Special Section 139D

Acknowledgement

This research work was supported by the 2014 KMITL grant.

Evaluation of the Efficiency of Selective Collection in a Small Town on the State of Rio Grande do Sul – Brazil

V E Schneider[1], M Poletto[1], D Peresin[1], S H Z Carra[1], D Vanni[1]

1 University of Caxias do Sul. Enviromental Sanitation Institute. Rio Grande do Sul State, Brazil. 95250-000, Getúlio Vargas Street, Caxias do Sul.

Abstract. With the increase of population concentration in urban areas, there is an increase in the solid waste generation, which demands the search for alternatives and solutions for the environmentally correct destination of these. In this context, this work presents an evaluation on the forms of organic and selective domestic waste collection and the potential for the recyclability of the waste destined to the same, based on the physical characterization and gravimetric composition of the solid wastes generated in the town of Antônio Prado, located in the state of Rio Grande do Sul, Brazil, between 2014 and 2016. It is observed that the population has significant information regarding the correct disposal of waste in the selective collection, since 60% of the waste destined to the same is effectively recyclable. Plastic (24.8%), paper (10.9%), glass (8.8%) and cardboard (8.4%) are the most representative materials in recycled waste samples in the urban area. The importance of continuity and improvement of environmental education programs is essential, due to the evolution in the quantity and complexity of products and materials currently manufactured, and to the method of mechanized waste collection used by the municipality.

1. Introduction

Modern society is constantly evolving and, as a result, the concentration of people in urban areas and the generation of solid waste is increasing. One of the current challenges is the search for alternatives and solutions for the final destination of these wastes, especially with regard to the prevention of soil, air and water pollution, since the inadequate management of these wastes is one of the causes of environmental, social, economic and health problems [1]. The fast growth of population, economy, consumption pattern, product obsolescence and new packaging accelerated the rate of urban solid waste generation, provoking a great challenge to their management [2], mainly at the municipal level.

For public managers around the world, the management of solid urban waste (MSW) has become, in the last decades, a matter of concern, due to the complexity of the materials and the quantity generated. According to the United Nations (UN), the current generation of waste in the world is around 12 billion tons/year and, by 2020, the estimated volume is 18 billion tons/year. A survey conducted by the Brazilian Association of Public Cleaning and Special Waste Companies shows that the volume of MSW generated in Brazil was 60 million tons/year in 2010, 6.8% higher than the previous year [3].

According to the National Solid Waste Policy [4], the Brazilian low about this subject, "selective collection and recycling are key instruments for the implementation of shared responsibility for the product life cycle, in addition to being fundamental to enable hierarchy in the solid waste management and the socioeconomic inclusion of scavengers". In this way, the Brazilian low directive establishes that the selective collection should be strongly considered in the plans, both federal and regional, and should have its viability encouraged.

According to [5] solid waste management must be offered by local public agencies, but it depends heavily on the active participation of the population that generates them. For this reason, environmental education is extremely important for the effectiveness of management. With it, it is likely that the rules of correct separation of residues are obeyed, materials that were previously discarded precipitously are reused and, even, there is a decrease in the generation of these materials.
The effectiveness of selective collection programs and initiatives necessarily requires the involvement of citizens and need for information and dissemination of programs and initiatives implemented. The community must be sensitized and motivated, and the concepts and practices must be assimilated and incorporated into the daily life of the population involved, with the purpose of ensuring its operationalization, viability and continuity, which are fundamental factors in achieving the expected results and durability [6].
Considering the need to evaluate the effectiveness of selective collection and community sensitization, this paper presents an evaluation on the forms of collection, the potential of solid wastes recyclability and the polymeric materials destined to the same, from the physical characterization and gravimetric composition of solid waste generated in the town of Antônio Prado, Rio Grande do Sul State - Brazil, between the years 2014 and 2016.

2. Methodology

The municipality of Antônio Prado has an estimated population of 13,296 inhabitants [7] and is located in the northeastern region of the Rio Grande do Sul State - Brazil. For the physical characterization and gravimetric composition of domestic solid waste generated in this town, a methodology based on the prerogatives of Brazilian Standard "ABNT NBR 10,007/ 2004" was used. Six samples were taken between December 2014 and November of 2016, covering urban and rural areas, mechanized collection and door-to-door collection, as well as selective and regular collection.
Solid waste samples were composed using three trucks: two compactors with capacity of 17 and 21 m³, and one with capacity of 14 m³, which operationalize the collection of waste in the town. The collection route followed specific routes in the areas defined by the municipal technical team, alternating streets and neighborhoods, so that the collected solid wastes were distributed homogeneously in the internal area of the truck. This procedure was adopted, because the town doesn't have a landfill making it impossible to discharge the waste collected from the truck. It should be noted that for this procedure the residues weren't compacted.
The residuals were collected directly from the truck's bucket for the composition of homogeneous and representative samples, stored in four 200 L barrels, totaling 800 L. Figure 1 shows the method adopted for the composition of the samples.

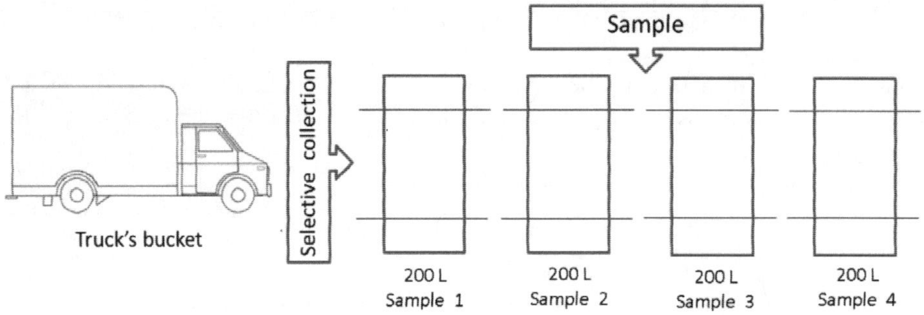

Figure 1. Methodology used to compose waste samples.

After the composition of the samples, the residues were segregated by type of material, weighted on a scale and grouped according to the treatment criteria defined by [8]:

• biodegradable: materials that can be reincorporated into biogeochemical cycles by decomposing organisms;
• recyclable: materials that can be reincorporated into industrial production cycles;
• disposable: materials for which there are still no processes that allow the return of their constituents to the natural or artificial cycles in a short time, or that their recycling is not economically viable.
For the per capita generation calculation of domestic solid waste destined to the selective collection, the information of resident population in the rural and urban area obtained in the 2010 population census [7], and the average of the weights of the months of July to November 2014, and from June to November 2016. For the identification of the percentages of materials found in the waste samples, the materials of the six characterization campaigns were added, the average of these values being calculated, also being calculated the percentage in relation to the total of residues sampled.
For the identification of the percentages of materials found in the waste samples, the materials of the six characterization campaigns were added, the average of these values being calculated, also being calculated the percentage in relation to the total of residues sampled.

3. Results

The amount of solid waste generated in the town was obtained from data provided by the Municipal Department of Agriculture, Environment and Industrial Development of the Municipality of Antônio Prado, referring to the average weight of the trucks that carry out the selective collection, in the period from July to November of 2014 and June to November 2016. Figure 2 shows the monthly amounts of solid waste generated in the town, destined to the selective collection.

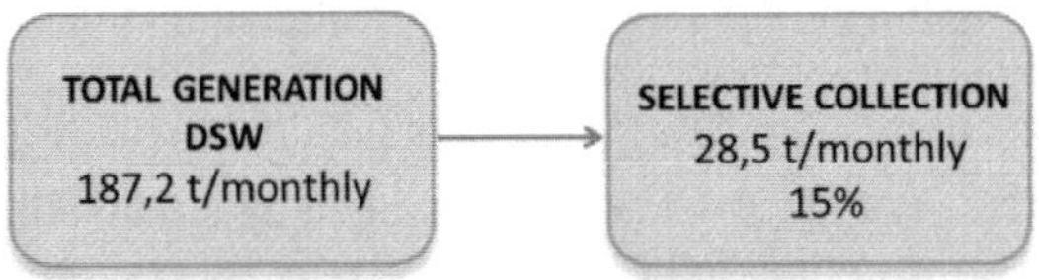

Figure 2. Monthly waste generation

As shown in Figure 3, 15% of the solid waste generated in the municipality of Antônio Prado is destined for the selective collection. It should be emphasized that in this quantification are included waste from the sorting center for recyclable waste that have no potential for recycling or are economically unviable to be recycled.
The per capita calculation of the solid waste sent to selective collection was 0.074 kg/hab.day (11.4% of the total generated in the town) of a total generation of 0.647 kg/hab.day of domestic waste. The total value obtained is in the range of values disclosed by [8], which establishes generations between 0.10 and 2.55 kg/hab.day, for municipalities with up to 30,000 inhabitants.
Figure 3 shows the percentage of representativeness of the categories of solid waste destined to the selective collection in the urban area (Figure 3 - A), which comprises the neighborhoods with door - to - door collection and the central area and neighborhoods with mechanized collection (containers) and in the rural area (Figure 3 - B).
The category of recyclables corresponded to 61.8% of the waste sampled from the selective collection in the urban area and 61.2% in the rural area. The biodegradable class accounted for 16.2% and 12.4%, and the disposable/waste class represented 22% and 26.4% respectively in urban and rural areas. It was assumed that the amount of biodegradable waste generated in the rural area and destined to the selective collection was smaller, since the use of this type of waste through composting is quite common in rural communities.
Considering the average percentage of the materials destined to the selective collection in the urban area, in the five campaigns carried out, it is observed that plastic (24.8%), paper (10.9%), glass (8.8%) and paperboard (8.4%) are the most representative materials in the samples of recyclable waste in the urban area. In the polymer material category, the highest percentages of PEBD - Film (6.2%), generally used in food packaging and PET - bottle (3.5%) were verified, represented by soft drink and water packages. It is observed that there is a tendency of the polymeric materials generated and

segregated by the population, to be destined to selective collection, regardless of whether they are rigid or film

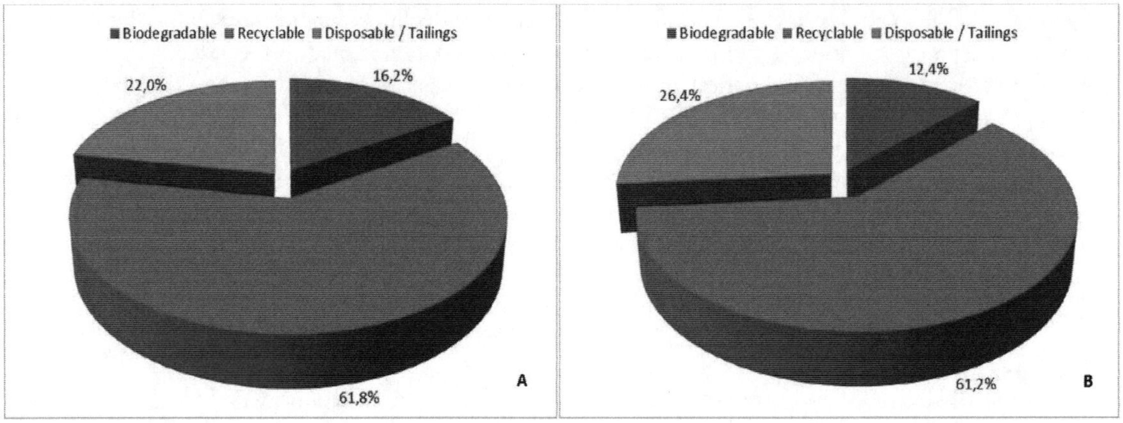

Figure 3. Composition of waste sent to the selective collection in the urban area.

Considering the representativeness of the potentially recyclable materials destined to the selective collection in the rural area, it is observed that the polymeric materials represent 42.4% of the total samples, being that of these materials, the ones found in greater quantity were: PP - Film (9, PEBD - Film (8.1%), consisting of food packaging, PP buckets and pots (4.3%) and PET bottles (3.3%). It should be noted that the "other category", contemplated in the category of polymeric material, refers to materials that have more than one type of polymer associated in its composition. The amount of polymer waste generated by the municipality is not enough for the implantation of a mechanical recycling plant of polymers, since the daily generation is less than 1000 kg per day for any type of polymer. However, the option of forming partnerships with other municipalities in the region through consortia can make mechanical recycling economically viable.

Figure 4 shows the composition of solid waste collected in the urban area destined to the selective, by containerized and door-to-door collection systems, grouped on the basis of the potential for treatability (biodegradable, recyclable and disposable). The containers (33 sets / recyclable + organic) were deployed in the central area of the town in March 2014. In July 2015, containerization was expanded to the neighborhoods around the central area, and 40 more (recyclable + organic), totaling 73 sets.

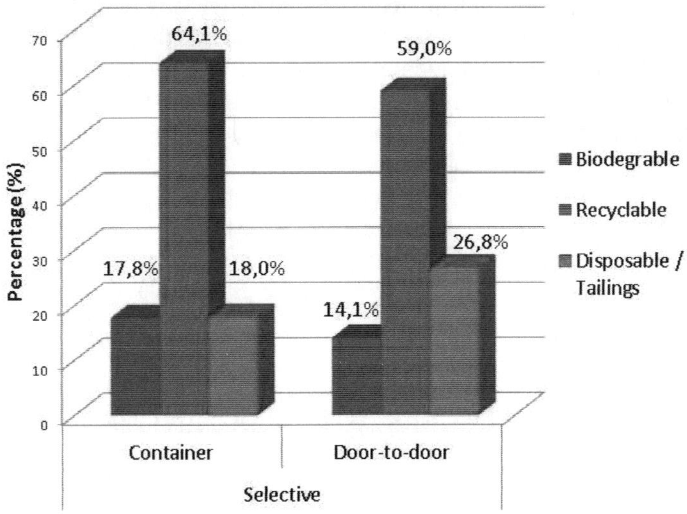

Figure 4. Potential of treatability of the waste destined to the selective collection, in the urban zone, by the systems of collected container and door-to-door.

Results show that the segregation of recyclables made by the population of the waste destined to the containers, although the difference is not significant, presents a trend of better segregation when compared to the door-to-door system.

In order to evaluate the efficiency of segregation and destination, after the expansion of the area occupied with containers, the percentage of waste destined to the collection was evaluated. It is observed that after the expansion of the containerization system in the neighborhoods, located in the environs of the central area, the waste destined to the containers of the selective collection, in the center, as in the districts, was more efficient, due to the greater percentage of residues for this collection.

It is also observed that both the amount of waste belonging to the biodegradable category and also the disposable solid waste was reduced by approximately 50% in the central area containers from 2014/2015 to 2016. This result is directly associated with the environmental education campaigns carried out by town. On the other hand, door-to-door collection in the neighborhoods had a reduction in the percentages of recyclables destined, which demonstrated a reduction in the efficiency of segregation and destination.

In door-to-door collection, one of the factors influencing the results is the collection times and days, since the trucks can collect at the wrong time or the population can place the waste on public roads at times and days that are not compatible with the calendar.

4. Conclusion

Considering the results obtained with six characterization campaigns of the town of Antônio Prado (December/2014, February/2015, June/2015, July/2016, September/2016, November/2016), it is concluded that, in general , the population has significant information regarding the correct disposal of solid waste in the selective collection, since 60% of the waste destined to it is effectively recyclable.

In the rural area of the municipality, which is attended by selective collection in 100% of its territory, consideration should be given to the implementation of incentive and environmental education programs for the practice of composting in rural communities. However, due to the presence of a representative quantity of biodegradable materials in the samples of the rural area, the need for more effective municipal programs for this practice is identified by the municipal management that the rural population performs the composting.

It is worth mentioning that the presence of biodegradable and disposable waste in the selective collection makes it difficult to segregate the sorting centers, generating odors, discomfort to the

segregators and the loss of the quality of the recyclable material, due to their contact and consequent contamination. With the presence of these materials in the plants, there is demand for transportation and final destination in landfills, resulting in increased costs in the waste management process.
The continuity and improvement of environmental education programs are essential, due to the evolution in the quantity and complexity of products and materials currently manufactured, and the relatively new method of collecting waste, adopted by the municipality (containers). In order to increase the efficiency of segregation, and consequently, in the management of domestic waste, it is indicated that the campaigns for the collection of recyclable materials, already developed by entities and public management in the municipality, are even more optimized, so that there is a greater dissemination of these information and greater adherence by the population.

5. References

[1] Oliveira, T.B., Galvão Junior, A.C. Municipal planning in the management of municipal solid waste and in the organization of the selective collection. Environmental Sanitation Engineering, [s.l.], v. 21, n. 1, p.55-64, 2016.
[2] Seo, S., Aramaki, T., Hwang, Y., Hanaki, K., 2004. Environmental impact of solid waste treatment methods in Korea. Journal of Environmental Engineering Div., ASCE 130 (1), 81–89. Gallardo, M.C., Peris, M., Colomer, F.J., 2014 Methodology to design a municipal solid waste generation and composition map: A case study. Waste Management. 34, 1920–1931.
[3] Dias, Sylmara Gonçalves. The Challenge of Urban Solid Waste Management, Society and Management.11(1), jan-jun/2012.
[4] Brasil. Federal low nº 7.404/2010. Regulates law n°. 12,305/2010, which establishes the National Solid Waste Policy, creates the Interministerial Committee of the National Solid Waste Policy and the Steering Committee for the Implementation of Reverse Logistics Systems, and provides other measures
[5] Miezah, K., Obiri-Danso, K., Kádár, Z., Fei-Baffoe, B., Mensah, M.Y. Municipal solid waste characterization and quantification as a measure towards effective waste management in Ghana. Waste Management, 46, 15-27, 2016.
[6] Bringhenti, J.R.,Günther, W.M.R. Social participation in selective urban solid waste collection programs. Environmental Sanitation Engineering, v. 16, n. 4, p.421-430, 2011.
[7] IBGE, Brazilian Institute of Geography and Statistics. Rio Grande do Sul. 2010.
[8] SCHNEIDER, V.E.; LIMA, L.M.Q. Characterization of domestic solid waste for defining management models – Case Study: Bento Gonçalves, Brazil. In: III Congreso Internacional de Quimica de La Anque, Puerto de la Cruz – Tenerife. p. 13-22. 1994.

IOP Conf. Series: Earth and Environmental Science **78** (2017) 012012 doi:10.1088/1755-1315/78/1/012012

Adsorption of Paraquat Dichloride by Graphitic Carbon Nitride Synthesized from Melamine Scraps

A Watcharenwong[1, 3*], A Kaeokan[1], R Rammaroeng[1], P Upama[1] and P Kajitvichyanukul[2]

1 School of Environmental Engineering, Institute of Engineering, Suranaree University of Technology, Nakhon Ratchasima, 30000, Thailand
2 Centre of Excellence on Environmental Research and Innovation, Faculty of Engineering, Naresuan University, 65000, Thailand
3 Center of Excellence in Advanced Functional Materials, Suranaree University of Technology Nakhon Ratchasima, 30000, Thailand.
Email: w.apichon@sut.ac.th

Abstract. In this research, graphitic carbon nitride (g-C_3N_4) was synthesized from useless melamine scraps. Mixture of melamine powder and urea was directly burned in the muffle furnace at 550 °C. Later as-synthesized g-C_3N_4 was modified with hydrochloric acid. The g-C_3N_4 powder was characterized by several techniques including X-ray diffraction, scanning electron microscope, and specific surface area analyser. Adsorption of the herbicide paraquat from an aqueous solution to suspended particles of g-C_3N_4 was investigated, taking into consideration several parameters such as initial concentration of paraquat, initial pH, and dosage of g-C_3N_4. The results showed that with the same amount of g-C_3N_4, the increase in the paraquat concentration caused the reduction in the removal efficiency and the higher the amount of g-C_3N_4, the less residual paraquat remained in the bulk solution. G-C_3N_4 showed better adsorption behaviour in the basic condition. Finally, Langmuir and Freundlich adsorption isotherms were also evaluated. Paraquat adsorption by g-C_3N_4 was in accordance with Langmuir more than Freundlich adsorption isotherm.

1. Introduction

Nowadays, environmental problem is a big issue worldwide. Industrial wastes from many industries had to manage appropriately. Big amount of investment was utilized for solid waste disposal facilities. For melamine tableware industry in Nakhon Ratchasima province, Thailand, huge amount of melamine scraps occurred each day of production. Moreover, the disposal cost of this waste as hazardous waste was rather expensive. Therefore the aim of this research was to utilize these non-valuable solid wastes.

Paraquat dichloride (1,1′-dimethyl-4,4′-bipyridinium dichloride) commonly referred to "paraquat," is one of the most widely used herbicides registered in Thailand. Paraquat is also often referred to Gramoxone (a common trade name). It is a widely utilized for the control of broadleaf weed in many agricultural and non-agricultural use sites. It is classified in non-selective herbicides. Recently, in the north of Thailand, farmers used paraquat 2-3 times/year in corn planting for animal food industry. Consequently, huge amount of paraquat remained in agricultural area. The reported half-life for paraquat in soil ranges from 16 months (aerobic laboratory conditions) to 13 years (field study) [1]. It affected to both animals and human. Paraquat exposures studies using frogs, birds and rodents at environmentally related concentrations, promote risks of declines in reproductive success due to embryotoxicity, delayed conception and malformations [2]. Accidental and intentional exposure of humans to paraquat still occurs even though its use is restricted. Brain damages and deaths can occurred in patients, who drank paraquat solution. Particularly, several studies suggest a link between paraquat use and Parkinson's disease (PD) [3]-[6]. Paraquat is highly soluble in water up to 620 g/L at 25 °C hence straightforwardly contamination into surface water by runoff. Conventional water supply unit process is unable to remove paraquat. Therefore, specific unit, for example, adsorption tank is required. From literatures review, graphitic carbon nitride (g-C_3N_4) was feasible adsorption material [7]. It can be synthesize from melamine directly. Consequently, this research will focus on synthesizing graphitic carbon nitride from useless melamine scraps and testing for paraquat adsorption.

2. Materials and Methods

2.1. Synthesis of Graphitic Carbon Nitride
The reagents used in this research were analytical grade. Melamine scraps was obtained from melamine tableware industry in Nakhon Ratchasima province, Thailand. The obtained melamine was crushed and sieved. In this experiment, melamine powder with the size of 100 mesh (0.149 -0.177 mm) was used. G-C_3N_4 was prepared by directly heating melamine with urea additive in the muffle furnace. The melamine to urea ratio was 10:0.3 g. The heating up rate was 10 °C per minute and after temperature reached 550 °C this temperature was holed for 4 hours. After it was naturally cooled to room temperature, g-C_3N_4 was obtained. The as-synthesized sample had been modified with 37% HCl for 3 hours. Then the modified sample had been washed with DI water for several time until the pH of water was 7. Finally, the sample had been dried in the oven at 105 °C for 3 hours. The finished sample was ground into powder before being used in the experiment.

2.2. Characterization of Graphitic Carbon Nitride
Powder X-ray diffraction (XRD) patterns of the g-C_3N_4 samples were recorded with an X-ray diffractometer model XD 6000. The morphology of g-C_3N_4 sample was observed on a scanning electron microscope (SEM) at 5kV for 30000X and 50000X magnification. The specific surface areas of the g-C_3N_4 sample was measured by nitrogen sorption on an autosorb iQ surface area analyzer model AUTOSORB and calculated by the Brunauer–Emmett–Teller (BET) method.

2.3. Adsorption Experiment
Paraquat adsorption experiment was conducted in batch type reactor. The experiment was divided into three parts. First experiment, paraquat at pH 7 was prepared in different initial concentration: 2, 6, and 8 mg/L and g-C_3N_4 of 2 g was used in each concentration. Second experiment, g-C_3N_4 with different amount of 0.4, 1, and 2 g was used in 8 mg/L paraquat at pH 7. Third experiment, 8 mg/L paraquat was prepared in different initial pH of 3, 7, and 10 and g-C_3N_4 of 2 g was used in each pH. All adsorption experiments were done in the dark. The adsorption experiments were conducted in a 1 L reactor containing 1 L paraquat solution with magnetic agitation. For each run of experiment, sample solution was taken off by a syringe at 5, 10, 15, 20, 25, 30, 45, 60, 90, 120, 180, 240, 300, and 360 minutes. Five mL turbid sample solution was filtered with 0.2 micron nylon filter. Then 1 % alkaline sodium dithionite solution of 1 mL was added to the clear solution. Shortly afterwards, the light blue color solution was obtained. Their respective concentrations were spectrophotometrically monitored at analytical wavelengths of 398 nm using a UV-VIS spectrophotometer.

3. Results and Discussion

3.1. Characterization of Graphitic Carbon Nitride
The XRD patterns for g-C_3N_4 sample was shown in Figure 1. The strongest XRD peak at 27.62°, was indexed for graphitic materials as (0 0 2) diffraction plane (JCPDS 87-1526) which is consistent with the reported g-C_3N_4 [8]. Figure 2 showed the SEM images of acid modified g-C_3N_4 particles at a magnification of 30,000x and 50,000x. The size of these nanoparticles was approximately 80 - 250 nm. G- C_3N_4 sample presented a BET specific surface area of 42.06 m^2/g (multi point mode measurement)

which was slightly less than other reports of 393.8 m^2/g for urea as precursor or 303.9 m^2/g for melamine as precursor [9]. However, its specific surface area was higher than natural clay (13.8 m^2/g) [10].

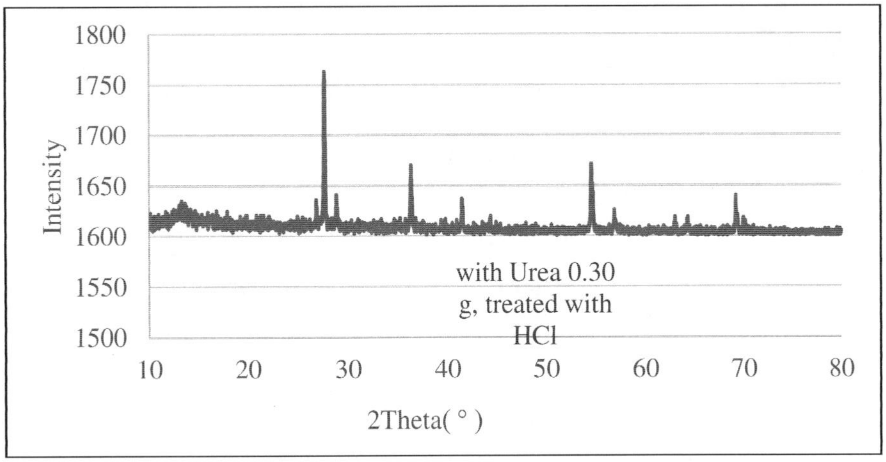

Figure 1. XRD patterns of g-C_3N_4.

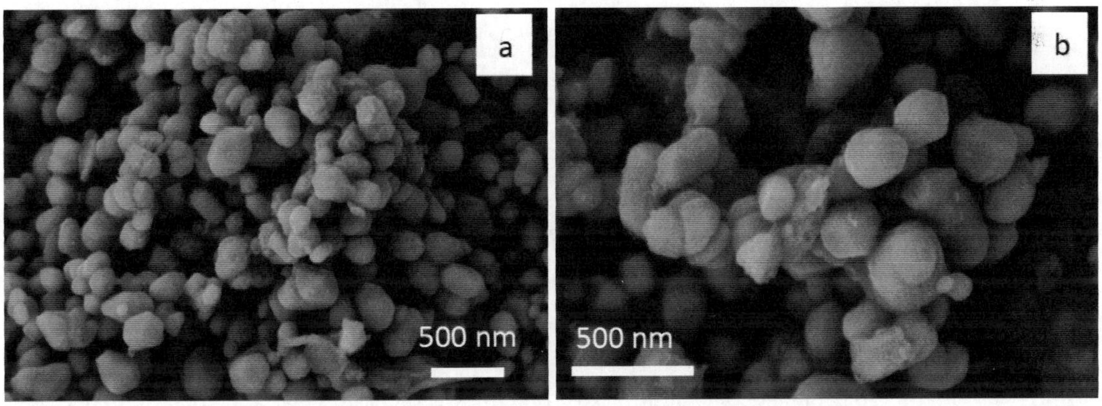

Figure 2. SEM images of g-C_3N_4 with a) 30000X and b) 50000X.

3.2. Paraquat Adsorption

From Figure 3 and Figure 4, the higher the amount of g-C_3N_4, the less residual paraquat remained in the bulk solution. This denoted the adsorption capacity of the g-C_3N_4 for a given concentration of paraquat. Considering the same initial quantity of paraquat, the increase in the adsorbent dosage caused an improvement in the removal efficiency leading to a greater amount of adsorbed paraquat, as expected.

IOP Conf. Series: Earth and Environmental Science **78** (2017) 012012 doi:10.1088/1755-1315/78/1/012012

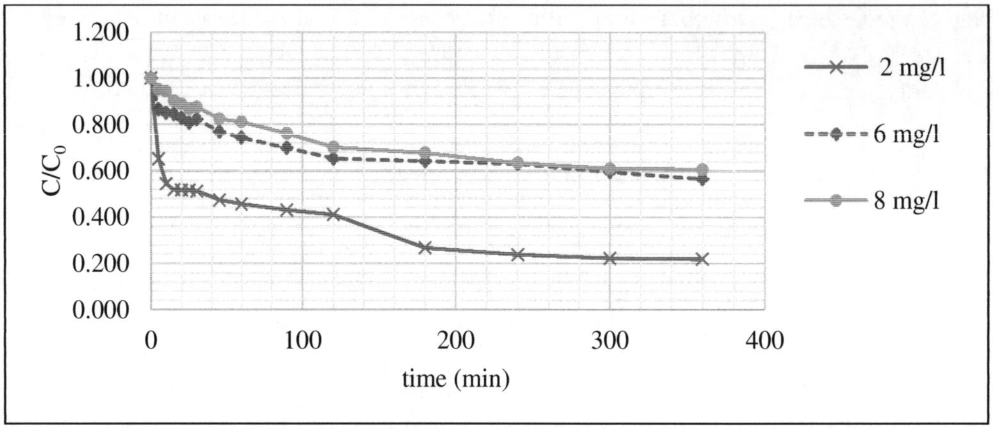

Figure 3. Variation in paraquat concentration overtime for g-C_3N_4 (2 g/L) suspensions exposed to a different concentration of paraquat solution pH 7 at room temperature.

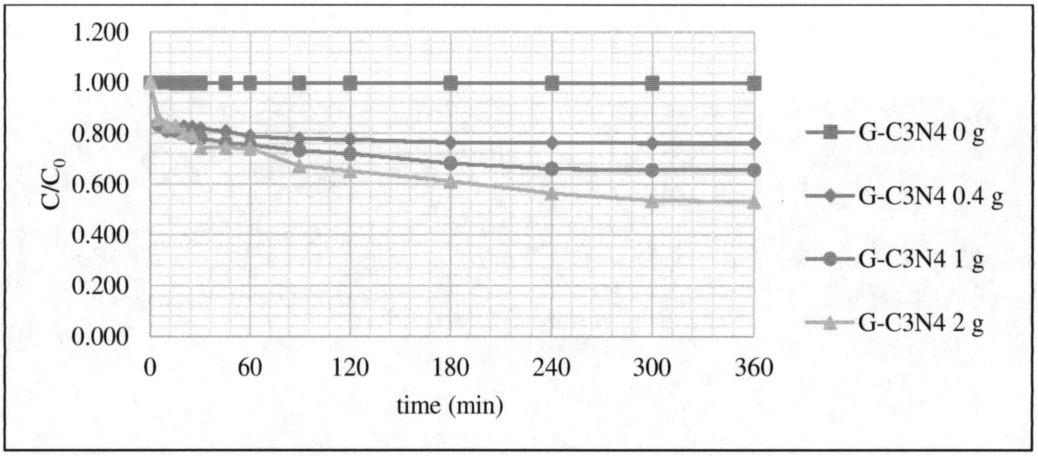

Figure 4. Variation in paraquat concentration overtime for different dosages of g-C_3N_4 (0, 0.4, 1, and 2 g/L) suspensions exposed to 8mg/L paraquat concentration solution pH 7 at room temperature.

From Figure 5, the effect of pH on adsorption capacity was evaluated using three different pH values (3, 7, and 10). The pH of solution was an important variable in the adsorption processes. The results showed that the adsorption behavior in basic condition was slightly higher than natural condition and obviously higher than acidic condition. Some researchers reported that the optimal pH range for the removal of paraquat was within 6–8, by also taking into consideration the realistic working pH conditions [11].

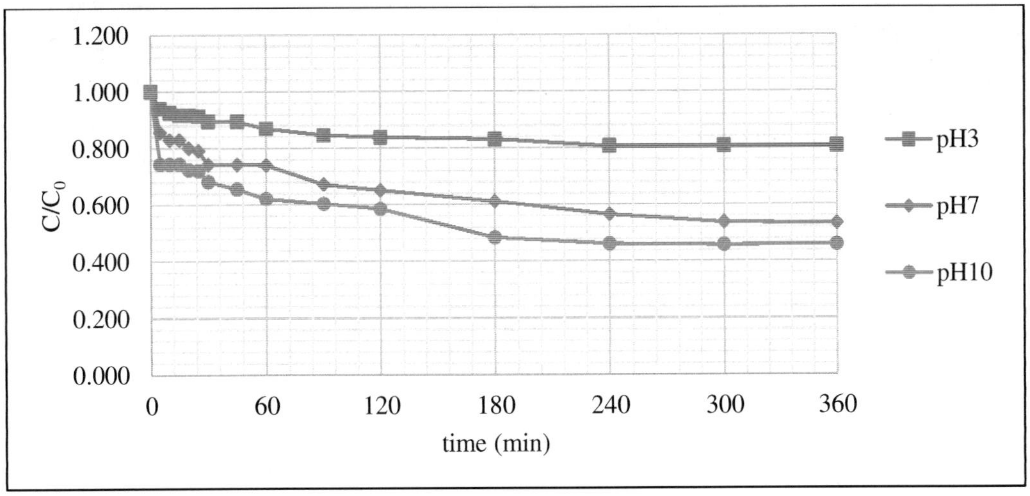

Figure 5. Variation in paraquat concentration overtime for different pH of 8mg/L paraquat solution with dosages of g-C_3N_4 2 g/L at room temperature.

3.3. Paraquat Adsorption Isotherms

Two adsorption isotherms were investigated, Langmuir and Freundlich isotherms. Both models (Langmuir and Freundlich) fit the experimental data quite well as in Figure 6. The correlation coefficient values (R^2) of these two isotherms were 0.9873 and 0.9623, respectively. The Langmuir isotherm presented slightly higher correlation coefficient values (R^2) which indicated that paraquat adsorption by g-C_3N_4 was in accordance with Langmuir adsorption isotherm. The Langmuir adsorption constant of 0.115 L/mg was calculated. This trend was in accordance with Hao's experiment [12] which studied the adsorption behavior of paraquat from aqueous solution using star fish particles.

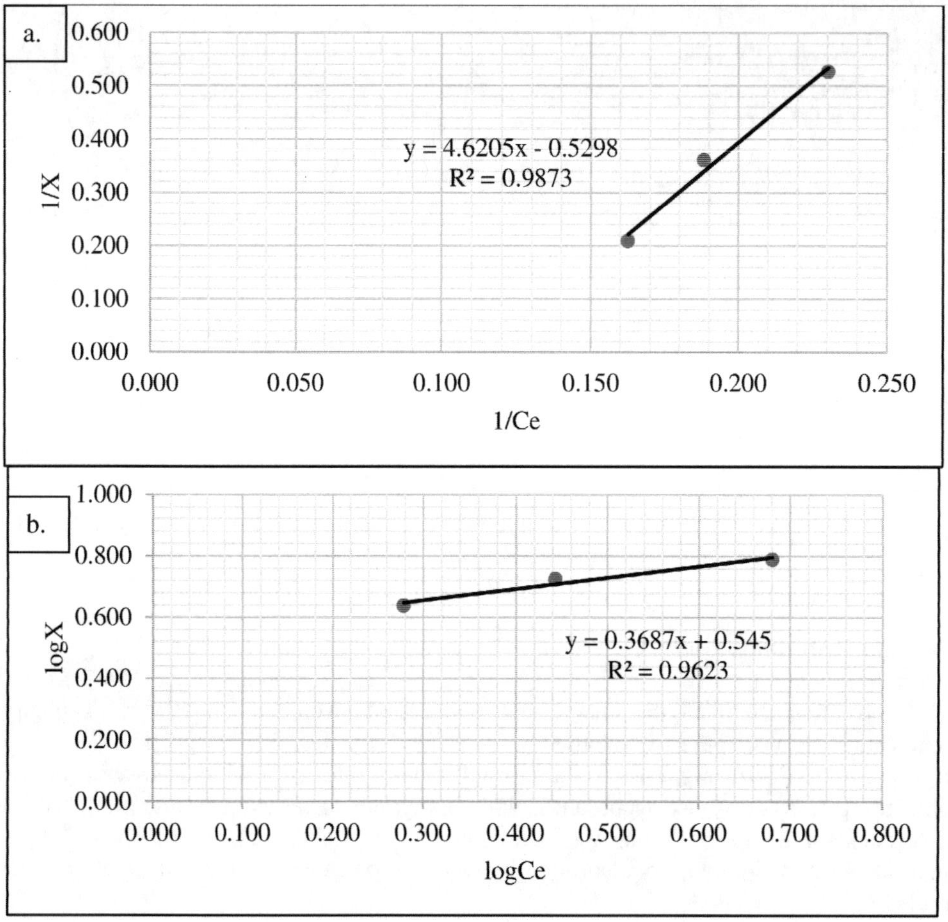

Figure 6. Adsorption isotherm of 8mg/L paraquat by different dosages of g-C_3N_4 (0.4, 1, and 2 g/L) at pH 7 a) Langmuir adsorption plot and b) Freundlich adsorption plot.

4. Conclusion

In summary, we have successfully fabricated the g-C_3N_4 material by directly heating the mixture of melamine and urea. The BET specific surface area of 42.06 m^2/g was achieved. Our results clearly indicated that the g-C_3N_4 has impartial performance in the paraquat adsorption. The higher the amount of g-C_3N_4 is, the higher adsorption capacity will be. It showed better adsorption behaviour in the basic condition. However, other important parameters such as particle size of adsorbent, contact time, stirring speed, and temperature should be examined in the future. Paraquat adsorption by g-C_3N_4 show good agreement to Langmuir more than Freundlich adsorption isotherm.

Acknowledgment

This work was funded by the Center of Excellence in Advanced Functional Materials, Suranaree University of Technology. Special thanks to Pimchanok Leuasoongnoen, Ariya Jindanant, Yotsapon Bailuang, and Narudon Saijaioup for supporting this research.

5. References

[1] Rao P S C, and Davidson J M 1980 *Environmental Impact of Nonpoint Source Pollution* eds. Overcash M R and Davidson J M (Ann Arbor Science) pp 23-67

[2] Hausburg M A, DeKrey G K, Salmen J J, Palic M R, and Gardiner C S 2005 Effects of paraquat on development of preimplantation embryos in vivo and in vitro *Reprod. Toxicol.* 20 pp 239–246.

[3] Moretto A, and Colosio C 2011 Biochemical and toxicological evidence of neurological effects of pesticides: The example of Parkinson's disease *NeuroToxicology.* 32 4 pp 383-391
[4] Andersen J K, and Di Monte D A 2014 Paraquat and Parkinson's Disease (PD), Reference Module in Neuroscience and Biobehavioral Psychology *Encyclopedia of the Neurological Sciences (Second Edition)* pp 792-794
[5] Pezzoli G, and Cereda E 2013 Exposure to pesticides or solvents and risk of Parkinson disease *Neurology* 80 pp 2035–2041
[6] Freire C, and Koifman S 2012 Pesticide exposure and Parkinson's disease: epidemiological evidence of association *Neurotoxicology* 33 pp 947–971
[7] Ye S, Qiu L G, Yuan Y P, Zhu Y J, Xia J, and Zhu J F 2013 Facile fabrication of magnetically separable graphitic carbon nitride photocatalysts with enhanced photocatalytic activity under visible light ***J. Mater. Chem. A* 1** pp 3008-15
[8] Ye L, Liu L, Jiang Z, Peng T, and Zan L 2013 Facets coupling of BiOBr-g-C_3N_4 composite photocatalyst for enhanced visible-light-driven photocatalytic activity *Appl. Catal. B: Environ.* 142–143 pp1–7
[9] Liu X, Wu X, Long Z, Zhang C, Ma Y, HaoX, Zhang H, and Pan C 2015 Photodegradation of Imidacloprid in Aqueous Solution by the Metal- Free Catalyst Graphitic Carbon Nitride using an Energy-Saving Lamp *Journal of Agricultural and Food Chemistry* 63 pp 4754−60.
[10] Dina A M, Manuel S, and Luís M 2015 Adsorption of paraquat dichloride to kaolin particles and to mixtures of kaolin and hematite particles in aqueous suspensions *Journal of Water Security* 1 pp 25-36
[11] Tiago F, Sofia F S, Tito T, and Ana L D 2017 Magnetic Hybrid Nanosorbents for the Uptake of Paraquat from Water. *Nanomaterials* 7(3), 68;doi:10.3390/nano7030068
[12] Hao H, Feng J, Chen W, Xiang S, Liu W, and Wu X 2015 Adsorption behavior of herbicide paraquat from aqueous solutions using starfish particles: kinetic, isotherm, and thermodynamic studies. *Asia-Pac. J. Chem. Eng.* 10 pp 347–355 DOI: 10.1002/apj.1872

Study on Spatial Model of Land Use Based on CA - Markov Model after Returning Cropland to Forest

Chun-xia QIU[1], Dong HAN[1], Qian-kun DONG[2] and Qin-qin MAO[1]

1 College of Geomatics, Xi'an University of Science and Technology, Xi'an, China
2 Shaanxi Institute of Surveying and Mapping of Land, Xi'an, China

Abstract. At present, the frontier of global environmental change research is land use / cover change, and an important way to understand global land change is to analyze and study typical areas. In order to study the law and the driving mechanism of land use change in the loess hilly and gully region, this paper takes the county south ditch watershed as an example, takes the remote sensing image of the basin in 1995, 2005 and 2015 as the data source, combined with the topographic data, And the spatial and temporal dynamics of land use before and after returning farmland to forest in Nangou watershed was analyzed. Finally, based on the land use situation in 2015, a variety of CA criteria were used to carry out modeling analysis to predict the land use in 2025 Type of purpose. And thus hope for the region to implement the scientific implementation of the project of returning farmland to forest.

1. Introduction

For a case study of the typical region. County in the south of the Loess Plateau gully region as the object of study, the state from 1974 on the watershed to plant trees and grass as the focus of the comprehensive treatment, [1]-[4] and in the years of comprehensive management and reduce sediment benefits The monitoring study has accumulated a lot of data and rich experience. [5] In order to analyze the spatial and temporal patterns and evolution of land use, the remote sensing image data and ground data were processed by remote sensing and GIS software, [6] and then the results were analyzed and compared. The temporal and spatial pattern of land use change and the evolution of land use change in Nangou watershed, and the quantitative simulation of land use change in 2025 based on CA-Markov model.

2. Survey Area Overview and Data Processing

2.1. *Overview of the Study Area*

County South ditch watershed (E 109 ° 19'30 ", N 36 ° 51'30") is located in the second sub-region of the hilly and gully region of the Loess Plateau. [7] It is a tributary of the lower reaches of the Xingzi River, which belongs to the typical loess hilly gully Area, belonging to Ansai County, Shaanxi Province, the basin area of 8.27 km², most of the soil is yellow soil, accounting for 77.1% of the total area. In the basin, the density of the gully in the terrain is 8.06 km/km², and the elevation is 1100-1400 m. [8]

2.2. The Specific Technical Process (it is shown in Figure 1)

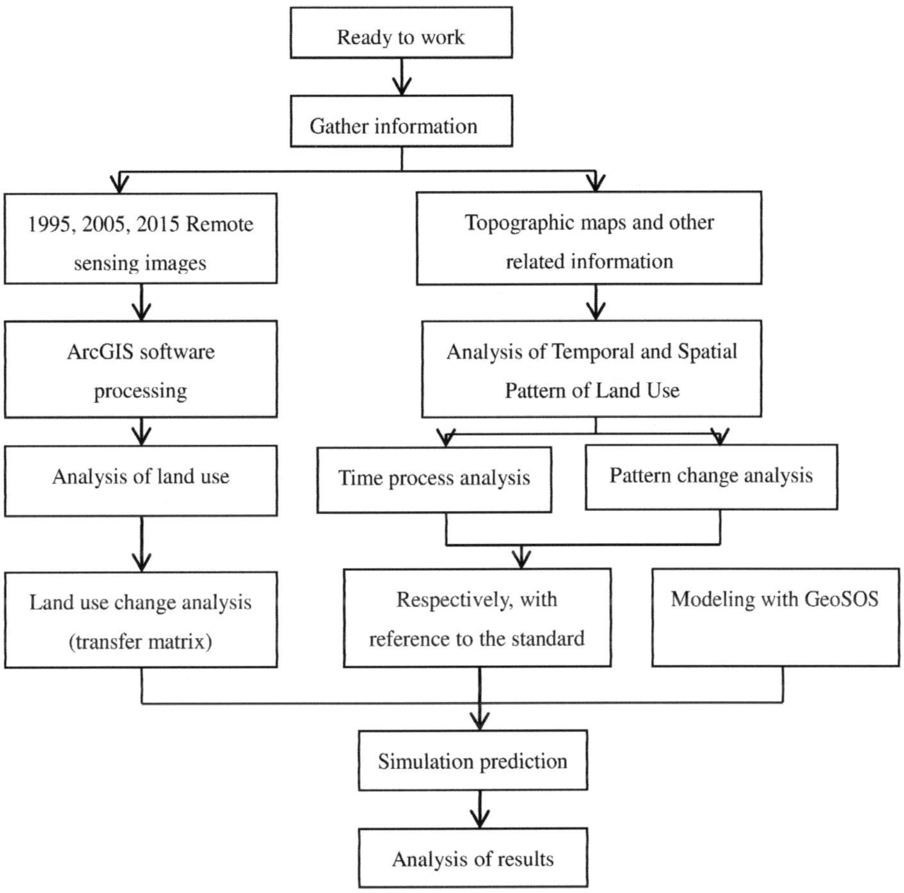

Figure 1. Technical flow chart

2.3. Data Pre-Processing

(1) ArcGIS data preparation;
(2) Data fusion;
(3) Overlay analysis;
(4) Calculate the area and export the attribute table;
(5) Export the data processing results;
(6) Data analysis.
① Time characteristics of land use structure.
② Land use type evolution analysis.

3. Temporal and Spatial Evolution of Land Use

3.1. Analysis of Land Use Time Process

3.1.1. Analysis of land use time before returning to cropland (1995-2005)
The relative rates of land use in the Nangou watershed in Ansai County from 1995 to 2005 are calculated, and as shown in Table 1.

Table 1. Relative change rate of major land use in 1995-2005

Land use type	The total land area of all types in 1995(m²)	The total land area of all types in 2005(m²)	Added area(m²)	Relative change rate of land use(m²)
Rural settlements	43315.79904	64697.02214	26381.2231	1.23
The river surface	88370.35988	66989.13678	19225.5163	0.90
Reservoir water surface	10645.02185	25026.24495	32301.1152	2.25
Dry land	107744.0591	66362.83602	445620.9716	10.77
Orchard	120184.9693	241566.1924	15320.6541	0.13
woodland	97319.63124	72938.40814	16589.5028	0.68
Shrubbery	449902.8673	261284.0904	17256.1265	0.09
Natural grassland	1752439.4911	1762439.489	42056.2546	4.21
Saline Land	399141.8641	187760.641	20154.9861	0.10
Cultivated land	1226174.781	1076827.288	12071.3695	0.08

3.1.2. Analysis on Land Use Time Process after Returning Farmland (2005 - 2015)

The relative rates of land use in the Nangou watershed in Ansai County from 2005 to 2015 are as shown in Table 2.

Table 2. Relative change rate of major land use in 2005-2015

Land use type	The total land area of all types in 1995(m²)	The total land area of all types in 2005(m²)	Added area(m²)	Relative change rate of land use(m²)
Rural settlements	64697.02214	103472.5069	55365.92072	1.43
The river surface	66989.13678	71729.70324	59327.64578	12.51
Reservoir water surface	25026.24495	14953.5889	11107.01465	1.10

Dry land	66362.83602	339715.6097	24114.76353	0.09
Orchard	241566.1924	423872.5792	57114.165	0.31
woodland	72938.40814	1762613.315	7218.090882	0.00
Natural grassland	1598384.547	1500489.002	496133.4712	5.07
Artificial grass	164054.9426	71937.97897	61623.32893	0.67
Saline Land	187760.641	166871.0906	118547.2083	5.67

Note: "0" indicates that the number of land use changes in a certain area during a study period is 0, that is, the relative change rate is zero.

3.2. Analysis of Spatial Pattern of Land Use

3.2.1. Analysis on spatial pattern of land use before returning to cropland (1995-2005)

Transformation of Main Land Use Types in Nangou Watershed of Ansai County in 1995-2005. The results showed that the increase of rural settlements and orchards was mainly due to the transfer of forest land area, 0.2190 km² and 0.7854 km² respectively, and the probability of turning into the orchards was 24.31%, and the orchards were Part of the area out of grass and shrubs, the transfer probability was 19.93% and 14.86%. The increase in cultivated land area is mainly due to the transfer of dryland and sparse forest land area into 0.1153 km² and 0.0989 km², respectively. Overall, the transfer rate of rural settlements is 90%, the conversion rate of orchards is 88%, and the transfer rate of reservoirs and forest land is the lowest. The area of dryland, shrub and orchard was the most obvious. In the period from 1995 to 2005, 0.9896 km², 0.8631 km² and 0.7854 km² were transferred to dryland, shrub and orchard respectively. The transfer probabilities were 30.63%, 29.81% and 24.31%, respectively.

3.2.2. Analysis on Spatial Pattern of Land Use after Returning Farmland (2005 - 2015)

The transformation of the main land use types in the Nangou Watershed of Ansai County in 2005 and 2015 shows that the increase of the area of forest land is mainly due to the transfer of land area of other types of land and forest land to 0.8605 km² and 0.1415 km ², the probability of turning grass into the shrubbery is 45.63%, and the probability of shrub forest turning into forest land is 35.78%. The increase of cultivated land area is mainly due to the conversion of grassland and other types of land area , Transferred to two separate 0.41 km² and 0.3102 km² respectively. Overall, shrubs, arable land and dry land transfer rate of the highest, are 94%, orchards and forest land transfer rate followed by 86% and 83%, river water transfer rate of the lowest, 17% The Grassland, shrubs and cultivated land conversion between the most obvious, during the 1.0082 km² and 0.4104 km² of grassland were converted to shrubs and arable land.

4. Simulation and Prediction of Dynamic Evolution of Land Use Change

4.1. Forecast of Land Use Distribution in 2025 Based on CA-Markov Model

Based on the CA-Markov model, that is, the probability of land use change in Nangou watershed of Ansai County based on Markov model, combined with the change of land use type in 2015, the time interval is 10 years. Forest policy measures, the appropriate adjustment of its probability, through the CA model of space, that is, from the spatial prediction of land use change pattern.

4.1.1. The determination of the transfer probability

Year as a unit, the change of land use type is divided into a series of discrete processes, according to the average annual conversion rate of each type (after returning farmland to forest, that is, from 2005 to 2015, the land use type conversion area divided by the age Interval 13 to obtain an average annual conversion area of the original area of the percentage of the area) to determine the land use type unit transfer probability, the rural residential area into other types of transition probability as the first line, the river surface into other landscape types The probability of transition as the second row, and so on, the establishment of the transition probability matrix, the 2005-2015 time period to determine the initial transfer probability matrix, [9]-[11] and as shown in Table 3.

Table 3. Transfer probability matrix for land use types in the initial state

	Residential Area	Forest land	Grass	Orchard	Arable Land	Dry Land	Saline Land	Waters
Residential Area	0.0657	0	0	0	0.0094	0.0015	0	0
Forest land	0.0044	0.3723	0.0297	0.0003	0.0252	0.0152	0	0.0273
Grass	0.0004	0.0293	0.0151	0.0015	0.0111	0.0011	0.0013	0
Orchard	0.0007	0.0370	0.0051	0.0182	0	0.0159	0	0
Arable land	0.0003	0.0131	0.0054	0.0407	0.0089	0.0085	0	0
Dry land	0	0.0377	0.0114	0	0	0.0279	0	0
Saline Land	0	0.0249	0.0016	0.001	0.0009	0.0485	0.0486	0
Waters	0	0.0089	0.0018	0	0	0.0074	0	0.0093

4.1.2. Dynamic simulation prediction

Based on the survey of land resources and the data of soil and water conservation in Nangou watershed in recent years, the transfer probability of the initial state of Table 3 is adjusted appropriately. The probability of converting cultivated land into forest land is adjusted to 0.0051, the conversion between forest land is 0.0723, the probability of conversion of cultivated land to grassland is 0.0008, the probability of conversion of grassland to forest land is 0.0041, and the probability of

IOP Conf. Series: Earth and Environmental Science **78** (2017) 012013 doi:10.1088/1755-1315/78/1/012013

conversion between residential area and other land use type is adjusted to 0. and the remainder remains unchanged. The adjusted transition probability matrix is shown in Table 4.

Table 4. Adjusted land use type transfer probability matrix

	Residential Area	Forest land	Grass	Orchard	Arable Land	Dry Land	Saline Land	Waters
Residential Area	0.0657	0	0	0	0	0	0	0
Forest land	0	0.0723	0.0297	0.0003	0.0252	0.0152	0	0.0273
Grass	0	0.0041	0.0151	0.0015	0.0111	0.0011	0.0013	0
Orchard	0	0.0370	0.0051	0.0182	0	0.0159	0	0
Arable land	0	0.0051	0.0008	0.0407	0.0089	0.0085	0	0
Dry land	0	0.0377	0.0114	0	0	0.0279	0	0
Saline Land	0	0.0249	0.0016	0.0010	0.0009	0.0485	0.0486	0
Waters	0	0	0	0	0	0	0	0.0093

4.1.3. Run the CA-Markov model

Specify the base land cover image for the land use map for 2015, enter the land use area transfer probability matrix from 2005 to 2015, and then input the adaptive image set, and specify the number of cycles of the geospatial automata to be 10, The cellular automaton filter creates a weight factor with significant spatial meaning, and uses the weighting factor to act on adjacent grid cells to change the state of adjacent grids. In the solution, a 5 x 5 filter was used to obtain the land use forecast for 2025 in the county.

4.1.4. The number of area changes in the forecast

The transfer matrix of the main land use types in the county south ditch basin from 2005 to 2015 is calculated by the formula 1. As shown in Table 5.

$$S_{ij}^{t+1} = f(S_{ij}^{t}, \Omega_{ij}, T) \qquad (1)$$

Table 5. Main land use type transfer matrix for Nangou watershed

	Residential Area(km²)	Forest land(km²)	Grass (km²)	Orchard (km²)	Arable (km²)	Dry Land(km²)	Saline Land(km²)	Waters Land(km²)
Residential Area	553	0	0	1	79	13	0	142
Grass	162	10812	5578	549	4101	397	483	11
Orchard	23	1162	159	571	0	499	0	0
Arable land	19	846	350	2629	577	546	0	0
Dry land	0	325	98	0	0	241	0	0
Saline Land	0	609	38	24	21	0	1185	0
Waters	0	106	22	0	0	88	0	704
Total	947	30037	8435	3788	5877	2466	1668	857

4.2. Dynamic Prediction of Land Use Change

According to the forecast, the area of settlements, woodlands, grasslands, orchards, arable land, dry land, saline land and waters will reach 131954.7 m², 4571225.3 m², 690310.7 m², 497044.2 m², 508629.6 m², 275068.8 m², 108581.1 m² and 75275.2 m² Because of the absence of 2025 years of remote sensing image data, the results can not be tested. However, this prediction results are consistent with the overall trend of land use after returning farmland to forest (2005-2015), indicating that the model is reasonable and good forecast of future land use change. The results show that land use change will continue to maintain the trend of 2005-2015 during the 2015-2025 period, and the area of forest land, orchard, rural settlements and dry land will increase as a whole, In 2015, the net increase in woodland was 1567525.3 m², an increase of 52.2%, mainly from the transfer of cultivated land and grassland, while the area of cultivated land, saline land, grassland and waters continued to decrease, The net area decreased by 79070.4 m², a decrease of 13.5%, mainly due to the transfer of woodland and orchard area. However, with the deepening of the project of returning farmland to forest, this increase and decrease will be reduced, because the area of returning farmland to forest is limited, the forest area changes over time will become saturated, and cultivated land can not be reduced. So the final forest land, grassland, arable land and other areas will remain basically stable, in a dynamic balance.

5. Conclusions

Based on ArcGIS software and CA-Markov model, this paper analyzes the relationship between the land use efficiency, the land use dynamic degree, the land use type relative change rate, the land use transfer matrix data and the relevant indicators, and analyzes the implementation of the policy of returning farmland to forest before and after the implementation the time course and spatial pattern of land use change in Nangou watershed of Ansai County, and the CA-Markov model was used to simulate the change of land use structure in Nangou watershed in 2025. And thus hope for the region to implement the scientific implementation of the project of returning farmland to forest.

6. References

[1] Yu Xing-xiu, Yang Gui-shan. Status Quo and Problems of Land Use / Cover Change in China [J]. *Advances in Geography*, 2002, 21 (1): 51-57.

[2] Wang Cheng, Han Xin-hai. Extraction of Hydrological Features Based on DEM in ArcGIS Environment - A Case Study of Nangou Watershed in County [J]., 2007 (4): 178-180.

[3] Dong Qian-kun.Study on Spatial and Temporal Pattern of Land Use in Zhifanggou Watershed Based on GIS [J]. *Mining Survey*, 2015 (4): 59-64.

[4] LU Kai.Study on cellular automata and its establishment [D]. *Harbin University of Science and Technology*, 2007.

[5] YANG Ding-ding, LUO Cheng-de, GONG Yuan-bo, et al. Soil nutrient dynamics of forest and grass complex model in returning farmland to forestland [J]. *FORESTRY SCIENCE*, 2007, 43 (s1): 101-105.

[6] TOBLERWR, A computer movie simulating urban growth in the Detroit region[J]. *Economic Geography*,1970,46:234—240.

[7] BATTYM, XIEYC. From cells to cities[J].*Environment and Planning B*,1994,21:531—548.

[8] WHITER, ENGELENG. Cellular automation and fractal form: a cellular Modeling approach to the evolution of urban land-use patterns[J]. *Environment and Planning A*,1993,25:1175—1199.

[9] Wang Wanzhong (1983). Study on the Relationship between Rainfall Characteristics and Soil Loss in Loess Area. *Soil and Water Conservation Bulletin* (4), 7-13.

[10] Liu Changping, Li Xia, Chen Yimin, Qin Yan, Li Shaoying, and Chen Minghui (2009). Landscape expansion index and its application in urban expansion analysis. *Acta Geographica Sinica*, 64 (12), 1430-1438.

[11] Wischmeier W H,Smith D D. Predicting Rainfall Erosion los-ses. A Guide to Conservation Planning with the Universal Soil Loss Equation (A) Agriculture Handbook No. 537, United States Department of Agriculture, Spring field,USA,1978.

Erosion of *Brassica incana* Genetic Resources: Causes and Effects

A Muscolo, G Settineri*, C Mallamaci, T Papalia and M Sidari

Agriculture Department, Mediterranea University Feo di Vito, 89124 Reggio Calabria, Italy
Email: amuscolo@unirc.it

Abstract: *Brassica incana* Ten., possessing a number of useful agronomic traits, represents a precious genetic resource to be used in plant breeding programs to broaden the genetic base in most *Brassica* crop species. *B. incana* that grows on limestone cliffs is at risk of genetic erosion for environmental constraints and human activities. We studied the pedological conditions of a Calabrian site where the *B. incana* grows, and we correlated the soil properties to the physiological and biochemical aspects of *B. incana* to identify the causes and effects of the genetic erosion of this species. Our results evidenced that physical soil conditions did not affect *B. incana* growth and nutraceutical properties; conversely, biological soil properties modified its properties. We identified leaf pigments and secondary metabolites that can be used routinely as early warning indicators of plant threat, to evaluate in a short term the dynamic behavior of plants leading to species extinction.

1. Introduction

The decline of biodiversity worldwide can lead to declines in overall levels of ecosystem functioning [1]. In the European context, there is a pressing need to assess the overall conservation status of species and habitats of community interest [2]. Rare species may be at greater risk of extinction because of their small geographic ranges, low abundances, and greater susceptibility to environmental changes [3], Further, incomplete information on their distributions, often gathered over long periods of time and with limited spatial accuracy, makes the assessment of these species particularly challenging [4], [5]. The genus *Brassica* is the most economically important genus within the Brassicaceae family being widely used in human diet as an important source of vegetables, condiments, and edible oils. The interest in Brassica vegetable crops has recently grown due to the breeding programs carried out in several countries - mainly in Asia, Europe and USA. Within *Brassica* species, an important loss of natural genetic diversity has been observed due to the introduction of highly performing F1 hybrids, characterized by a narrow genetic base that is contributing to the slow disappearing of landraces. In the last decades, an important loss of natural genetic diversity (loss of wild germoplasm) has also been caused by climatic changes, agricultural practices, deforestation and human activities. Calabria represents one of the main diversification center of the genus *Brassica.* The species most distributed in Calabria are *B. fructicolosa, incana, napus, rupestris, gravinae, nigra, montana and oleracea* [6]. These populations are wild relatives of kale crops [7] and therefore represent an important genetic resource. The populations are characterized by a considerable morphological variation [8] and show a strict genetic relationship [9]. Among these populations, *B. incana,* native to east and southeastern Europe (USDA, ARS, National Genetic Resources Program *2010, Marhold 2011),* possesses a number of useful agricultural traits. Additionally, from a nutraceutical point of view, it contains high amounts of glucosinolates and phenols, compounds with antioxidant, anticarcinogenic, cardiovascular and biocide properties that confer to *B. incana,* phytotherapic properties, encouraging its use in medicine, and in organic farming. Additionally, seeds of *B. incana* are a source of mustard powder as *B. juncea* and *B. nigra*, and could represent a benefit for the local economy. Many International Institutions, involved in the nature protection, listed this species in the "IUCN Red List of Threatened Species" that deserve attention and conservation priority. As a conservation tool, red list data are recommended to be used at various scales, including site level evaluations and national resource management and legislation [10], [11]. At the local level, the presence of species recognized as threatened by an authoritative system can be accurate pointers for prioritizing key habitats and their conservation [12], [13]. *B. incana* is assessed as Data Deficient as there are currently insufficient information available to evaluate this species, thus researches are needed to determine the current demographic status of this species. Based on this background, the goal of this work was to correlate the soil properties of the Calabrian sites where the *B. incana* grows to the physiological and biochemical aspects of *B. incana* to identify the causes and effects of the genetic erosion of this species.

2. Materials and methods

The soils were collected under *B. incana* vegetation in Stilo, Calabria, South Italy, (Lat. 38° 30' 29,2'' N, Long. 16° 22' 45,8'' E, altitude of 600 m above sea level, with a mean annual air temperature of 15.6 °C and a mean annual rainfall of 905 mm) from 2 different layers (0-10; 10-20 cm) separated on the basis of morphological differences which could be perceived by the naked eye. Soil samples (1 kg) were taken from each layer and analyzed separately. The samples were brought to the laboratory on the same day. Prior to the soil analysis, except for dehydrogenase (DH) activity and fluorescein diacetate hydrolysis (FDA), all the soil samples were air-dried, sieved (<2mm), and visible roots were removed.

Particle size analysis was detected following Bouyoucos [14] method; pH was measured in distilled water and in 1 M KCl using a 1:2.5 (soil:water) suspension; organic carbon (OC) was determined by dichromate oxidation, following the method of Walkley and Black[15]; soil total nitrogen (TN) was measured by the method of Kjeldahl [16]; electric soil conductivity (EC) was measured in distilled water using a 1:2.5 (soil:water) suspension by using a conductometer. Phenols were extracted with distilled water, 1:10 (w/v) [17]. Total water-soluble phenols (WSP) were measured by using the Folin–Ciocalteau reagent, following the Box method [18]. Tannic acid was used as a standard and the concentration of water-soluble phenols was expressed as tannic acid equivalents (mg TAE/g dw) [19]. FDA hydrolysis reaction was determined according to the methods of Adam and Duncan [20]. Dehydrogenase (DH) activity was determined by the method of von Mersi and Schinner [21].

Leaves of *B. incana* were homogenized in a chilled mortar with distilled water at a ratio of 1:4 (leaves/water; w/v) and centrifuged at 14000 g for 30 min. All steps were performed at 4 °C. The supernatants were filtered through two layers of muslin cloth and were used to determine the total antioxidant capacity by the spectrophotometric method of Prieto et al. [22]. Chlorophyll a and b content, expressed in μg/g fresh weight, was determined following the method of [23]. Total phenolic content was determined with the Folin–Ciocalteu reagent according to a modified procedure described by Singleton and Rossi [24]. Carotenoid content was determined according Lichtenthaler [25].

All variables analyzed were first tested for data normality by using Shapiro- Wilk's test. All data were analyzed by one-way analysis of variance (ANOVA) followed by Tukey's multiple comparison test at a 95% confidence level. All data collected were statistically analyzed using SYSTAT 8.0 software.

3. Results

Tables 1 summarizes physical, chemical and biochemical properties of the soils collected under *B. incana* vegetation.

Table 1. Physical chemical and biochemical properties of soil under B. incana vegetation: electric soil conductivity (EC, dS /m); organic carbon (OC %); organic matter (OM %); total nitrogen (TN %); carbon/nitrogen ratio (C/N); water-soluble phenols (WSP, mg TAE/g dw); Dehydrogenase (DH, μg TFF g-1h-1); Fluorescein diacetate (FDA μg fluorescein g-1h-1). Different letters in the same column indicate significant differences $p \leq 0.05$.

Depth (cm)	Sand %	Loam %	Clay %	pH (H_2O)	pH (KCl)	EC	OC	OM	TN	C/N	WSP	DH	FDA
0-10	85	10	5	7.94^b	6.86^b	< 1	0.350^a	0.602^a	0.045^a	7.8^b	120.51^a	17.39^a	85^a
10-20	86	9	5	8.12^a	7.31^a	< 1	0.130^b	0.224^b	0.015^b	8.6^a	105.63^a	9.78^b	64^b

pH was lower at the soil surface than at the depth. pH (H_2O) was slightly alkaline ranging from 7.94 to 8.12. pH (KCl) ranged from 6.86 to 7.31, and it was consistently less than that measured in H_2O. Soil had a sandy texture, an EC lower than 1 dS/m, and a low content of organic matter and total nitrogen. The C/N ratio increased with increasing soil depth. DH and FDA hydrolysis activities were higher in the surface layers. The greatest amount of WSP was detected in the surface layer. The plants measured, in mean, 40 cm (data not shown). The amount of total chlorophyll and chlorophyll *a* was low as well the Chlorophyll *a* and *b* ratio and carotenoids.

Table 2. Content of antioxidants, chlorophylls, carotenoids and total phenols in leaves of *B. incana*.

Anthocyanins µg anthocyanin gr^{-1} FW	Chlorophyll *a* mg 100 gr^{-1}FW	Chlorophyll b mg 100 gr^{-1}FW	Chl	Chla/Chl b	Carotenoids mg gr^{-1} FW	Total Phenols µg TAET g^{-1} FW	Total Antioxidant capacity (µg tocopherol ml^{-1})
2.4	94.7	60.6	117.2	1.56	8.0	4342	462.0

Conversely a great amount of phenols was detected in leaves of *B. incana* the total antioxidant capacity was high too (Table 2).

4. Discussion

The loss of genetic variability represents not only the loss of wild germplasm but also the loss of evolved landraces resulting from the interaction of environmental selection with the genes present in both wild and cultivated populations [26]. The *Brassica* genus has not been an exception and, in particular, conservation of wild *Brassica* species has been a high priority [27]. During the 1970s, wild germplasm of *Brassica* was extensively collected and cytogenetic studies were started. Intensive efforts were made in the last decades to search and collect this material that, otherwise, would be irreversibly lost [28], [29]. After 1970s, the introduction of the concept of biodiversity was a strong support for many improvements in ex situ and in situ conservation strategies. *Brassica* wild relatives are valuable sources of genes for crop improvement [27]. They have been widely used since more than one century to increase resistance/tolerance to biotic and abiotic stresses in several cvs. Recently they have been used to increase nutritional value in edible parts like enhancing proteins or vitamins content. Conservation of species involves the selection of accessions to be conserved and the maintenance of these accessions for current and future users by regeneration. Decisions concerning both these aspects require specific knowledge of the environment, and soil characteristics where the species naturally grows and also physiological and biochemical knowledge of the species. Our results evidenced that *B. incana* was less high than the Sicilian *B. incana* [27] suggesting that is spread in sandy eroded soils with low amounts of organic matter not for choice but almost forced by climatic changes and human interventions that allowed other plants to take over confining this species to marginal places where other plants difficulty grow and/or animals hardly access to use this species as fodder. Thus, *B. incana* used these sites as shelters for own survival and conservation. Data of biochemical traits of this accessions supported this hypothesis. Plant secondary metabolites and pigment synthesis display an important role in the adaptation of plants to their environment [30], and generally low pigment synthesis has been proven to be the consequence of hexogen stress, senescence or adaption to changing environments [31]. In this sense, plants to maintain a balanced physiological state in the respective tissues conferring protection against environmental stresses [32]-[34] use chlorophylls, carotenoids, anthocyanins, and phenols. Our data evidenced that *B. incana* had a lower content of anthocyanins, chlorophylls, carotenoids, and total phenols in respect to data reported in literature [35]. All these data evidence that the properties of the soil in situ is not optimal for *B. incana* growth. Soil under brassica with a low amount of organic matter, nitrogen, FDA and DH, enzymes related to soil quality, induced low production of pigments and antioxidant compounds, affecting health-properties in *B. incana.* A negative linear relationship between the amount of photosynthetic pigments, antioxidant compounds, organic matter and quality of soils and a positive linear relation between leaf phenols and amount of organic matter, and soil enzyme activities was observed. In short, these data demonstrate the low fertility of soil in terms of organic matter and the scarce soil quality in terms of enzymatic activity and soil ecosystem functioning are the cause of genetic erosion risk and scarce propagation of *B. incana* [36], [37].

5. Conclusion

Results from current study show that decrease in soil fertility (chemical and biological soil characteristics), caused by human interventions or environmental changes, can alter species distribution in a geographic area reducing plant its spread and causing erosion genetic risk. Additionally, soil biological properties can condition plant growth and metabolism, affecting the nutraceutical properties of a vegetal species. The results of this study support the idea of using early key leaf biochemical indicators (decrease in pigments and contemporary increase in total phenols) for investigating environmental constraints inducing species extinction, specifically in the case of brassicacae. Despite this progress, several areas of uncertainty remain to be investigated.

6. Acknowledgments

The authors gratefully acknowledge Prof. Giovanni Spampinato of Agriculture Department for his valuable support in the identification of the species. This research was supported by Dipartimento 6 Agricoltura, Foreste e Forestazione, della Regione Calabria, Programma di sviluppo Rurale 2007/2013 misura 214, AZIONE 6 " Salvaguardia e recupero della *Brassica incana* in Calabria".

7. References

[1] Rands M R W, Adams W M, Bennun L, Butchart S H M, Clements A, Coomes D, Entwistle A, Hodge I, Kapos V, Scharlemann J P W, Sutherland W J and Vira B 2010 Biodiversity conservation: Challenges beyond 2010 *Science* 39 1298-1303

[2] European Commission, Council Directive 92/43/EEC of 21 May 1992 on the conservation of natural habitats and of wild fauna and flora, 1992L0043 Brussels, Belgium: The Council of the European Community

[3] Pimm S L, Jenkins C N, Abell R, Brooks T M, Gittleman J L, Joppa L N, Raven P H, Roberts C M and Sexton J O 2014 The biodiversity of species and their rates of extinction, distribution, and protection *Science* 344 6187

[4] Gogol-Prokurat M 2011 Predicting habitat suitability for rare plants at local spatial scales using a species distribution model *Ecol. Appl.* 21 33-47

[5] Marcer A, Sáez L, Molowny-Horas R, Pons X and Pino J 2013 Using species distribution modelling to disentangle realised versus potential distributions for rare species conservation *Biol. Conserv.* 166 221-230

[6] Raimondo F M, Mazzola P, Ottonello D 1991 On the taxonomy and distribution of *Brassica* sect. *Brassica* (Cruciferae) in Sicily *Flora Med.* 1 63-86

[7] Heywood V H and Zohary D 1995 A catalogue of wild relatives of cultivated plants native to Europe *Flora Med.* 5 375-415

[8] Raimondo F M and Mazzola P 1997 A new taxonomic arrangement in the Sicilian member of *Brassica L.* sect. *Brassica Lagascalia* 19 831-838

[9] Geraci A, Chèvre A M, Divaret I, Eber F and Raimondo F M 2004 Isozyme analysis of genetic diversity in wild Sicilian populations of *Brassica* sect. *Brassica* in view of genetic resources management *Gen. Res. Crop. Evol.* 51 137-146
[10] Rodríguez-Sánchez F, Pérez-Barrales R, Ojeda F, Vargas P and Arroyo J 2008 The strait of Gibraltar as a melting pot for plant biodiversity *Quat. Sci. Rev.* 27 2100-2117
[11] IUCN 2011 Guidelines for Using the IUCN Red List Categories and Criteria. Version 10.1. Prepared by the Standards and Petitions Subcommittee in September 2011
[12] Meynell P J 2005 Use of IUCN Red Listing process as a basis for assessing biodiversity threats and impacts in environmental impact assessment *Impact Assess. Proj. Apprais.* 23 65-72
[13] Batary P, Baldi A and Erdos S 2007 The effects of using different species conservation priority lists on the evaluation of habitat importance within Hungarian grasslands *Bird Conserv. Int.* 17 35-43
[14] Bouyoucos G J 1962 Hydrometer method improved for making particle-size analyses of soils *Agron J* 54 464-465
[15] Walkley A and Black I A 1934 An examination of the Degtjareff method for determining soil organic matter and a proposed modification of the chromic acid titration method *Soil Sci* 37 29-38
[16] Kjeldahl J 1883 Neue methode zurestimmung des stickstoffs in organischen korpen *Z. Anal. Chem.* 22 366-382
[17] Kaminsky R and Muller W H 1978 A recommendation against the use of alkaline soil extractions in the study of allelopathy *Plant Soil* 49 641-645
[18] Box J D 1983 Investigation of the Folin–Ciocalteau reagent for the determination of polyphenolic substances in natural waters *Water Res.* 17 511-525
[19] Kuiters A T and Denneman C A J 1987 Water-soluble phenolic substances in soils under several coniferous and deciduous tree species *Soil Biol. Biochem.* 19 765-769
[20] Adam G and Duncan H 2001 Development of a sensitive and rapid method for the measurement of total microbial activity using fluorescein diacetate (FDA) in a range of soils *Soil Biol. Biochem.* 33 943-951
[21] von Mersi W and Schinner F 1991 An improved and accurate method for determining the dehydrogenase activity of soils with iodonitrotetrazolium chloride *Biol. Fert. Soils* 11 216-220
[22] Prieto P, Pineda M and Aguilarm M 1999 Spectrophotometric quantitation of antioxidant capacity through the formation of a phosphomolybdenum complex: specific application to the determination of vitamin E *Anal. Biochem.* 269 337-341
[23] Arnon D I 1949 Copper enzymes in isolated chloroplasts, polyphenoxidase in *Beta vulgaris* *Plant Physiol.* 24 1-15
[24] Singleton V L and Rossi J 1965 Colorimetric of total phenolics with phosphomolybidc-phosphotungstic acid reagents *Am. J. Enol. Vitic.* 16 144-158
[25] Lichtenthaler H K and Wellburn A R 1983 Determinations of total carotenoids and chlorophylls a and b of leaf extracts in different solvents *Biochem. Soc. Trans.* 11 591-592
[26] Mercer K L and Perales H R 2010 Evolutionary response of landraces to climate change in centers of crop diversity *Evol. Appl.* 3 480-493
[27] Branca F and Cartea E 2011 Brassica *Wild Crop Relatives: Genomic and Breeding Resources Oilseeds* (Berlin Heidelberg: Springer-Verlag) ed C Kole chapter 2 pp 17-36
[28] Gómez-Campo C 2006 Erosion of genetic resources within seed genebanks: the role of seed containers *Seed Sci. Res.* 16 291-294
[29] Gómez-Campo C, Aguinagalde I, Arús P, Jiménez-Aguilar C, Lázaro A, Martín-Clemente JP, Parra-Quijano M, Sánchez-Yélamo MD, Simonetti E, Torres E, Torcal L and Tortosa ME 2007 Geographical distribution and conservation status of *Brassica* montana in NE Spain *Cruciferae News* 127 32-34
[30] Alvarez M A 2014 *Plant Biotechnology for Health: from Secondary Metabolites to Molecular Farming* (Cham: Springer International Publisher) DOI. 10.1007/978-3-319-05771-2_3 ebook
[31] Gould K S, Kuhn D N, Lee D W and Oberbauer S F 1995 Why leaves are sometimes red *Nature* 378 241-242

[32] Grassmann J, Hippeli S and Elstner E F 2002 Plant's defence and its benefits for animals and medicine: role of phenolics and terpenoids in avoiding oxygen stress *Plant Physiol. Biochem.* 40 471-478
[33] Mittler R 2002 Oxidative stress, antioxidants and stress tolerance *Trends Plant Sci.* 7 405-410
[34] Blokhina O, Violainen E and Fagerstedt K V 2003 Antioxidants, oxidative damage and oxygen deprivation stress: a review *Ann. Bot.* 91 179-194
[35] Maxted N, Iriondo J M, Hond L, de Dulloo E, Lefèvre F, Asdal A, Kell S P and Guarino L 2008 *Conserving plant genetic diversity in protected areas: population management of crop wild relatives* ed J M Iriondo, N Maxted *et al* (Wallingford: CABI) p 288 ISBN 9781845932824 DOI 10.1079/9781845932824.0000
[36] Choe J H, Kim H Y, Kim Y J, Yeo E J and Kim C J 2014 Antioxidant activity and phenolic content of persimmon peel extracted with different levels of ethanol *Int. J. Food Prop.* 17 1779-1790
[37] Wong J Y, Matanjun P, Ooi Y B H and Chia K F 2014 Evaluation of antioxidant activities in relation to total phenolics and flavonoids content of selected malaysian wild edible plants by multivariate analysis *Int. J. Food Prop.* 17 1763-1778

IOP Conf. Series: Earth and Environmental Science **78** (2017) 012015 doi:10.1088/1755-1315/78/1/012015

Review in Strengthening Technology for Phytoremediation of Soil Contaminated by Heavy Metals

Chishan Wu[1,2], Xingfeng Zhang[1,2] and Yang Deng[1,2]

1 Guangxi Key Laboratory of Environmental Pollution Control Theory and Technology, Guilin University of Technology, Guilin 541004, China
2 Guangxi Collaborative Innovation Center for Water Pollution Control and Water Safety in Karst Area, Guilin University of Technology, Guilin 541004, China
E-mail: 466420176@qq.com

Abstract. In view of current problems of phytoremediation technology, this paper summarizes research progress for phytoremediation technology of heavy metal contaminated soil. When the efficiency of phytoremediation may not meet the demand in practice of contaminated soil or water. Effective measures should be taken to improve the plant uptake and translocation. This paper focuses on strengthening technology mechanism, which can not only increase the biomass of plant and hyperaccumulators, but also enhance the tolerance and resistance to heavy metals, and application effect of phytoremediation, including agronomic methods, earthworm bioremediation and chemical induction technology. In the end of paper, deficiencies of each methods also be discussed, methods of strengthening technology for phytoremediation need further research.

1. Introduction

Recent years had witnessed the expanding area of soil contaminated by heavy metal,owing to rapidly increasing of population, the unreasonable development and utilization of mineral resources, the rapid development of pesticide and applying chemical fertilizers, sewage irrigation and industrialization and so on[1].Toxic heavy metal can poses a serious threat to human health, due to bioaccumulation in humans and it can not be easily degraded , so the remediation of heavy metal contaminated soil is particularly important.

Phytoremediation is an environmentally-friendly technology, using plants to remove heavy metal from contaminated soil[2]. It has advantages over physical and chemical remediation,such as permanent effect, cost-effective and secondary pollutant is easy to control. However, many factors limited the remediation efficiency in practice, such as :1)The majority of hyperaccumulating species discovered so far have low biomass ,restrict to region or have a long remediation cycle. 2) They might be good at hyperaccumulating one metal, but be sensitive to other contaminant. 3)The roots of hyperaccumulator plants are generally shallow, and the ability to repair the deep soil is limited.

Thus, it's a important direction to enhance the efficiency of phytoremediation by strengthening technology. This paper mainly discusses how to improve plant biomass and accumulation of heavy metals from physical, chemical and agronomic measures.

2. Agronomic Practices

2.1. Use of Fertilization

Fertilization can promote plant growth and biomass, and then increase the total amount of heavy metals accumulated in plants. It can change the form and activity of heavy metals, and then affect the absorption and accumulation of heavy metals in plants. In the study on fertilization of phytoremediation, nitrogen, phosphorus, potash, compound organic fertilizer and CO_2 gas fertilizer are the most commonly used.

Robinson[3] found that the application of nitrogen fertilizer could increase the biomass of Alyssumbertolonii ,which was nickel hyperaccumulator, by two times without decreasing its aboveground nickel content. However, the amount of fertilizer should be within a suitable range. Excessive application of fertilizer does not necessarily increase the accumulation of heavy metals in plants, but reduces the efficiency of phytoremediation[4]. Field trials conduced by Liao also showed that the soil remediation efficiency decreased from 7.84% to 6.63% when the phosphorus application rate was 340 kg/hm^2 and 600kg/hm^2.

Application of CO_2 fertilizer in phytoremediation, can not only enhance the resistance of plants to pollutions, improve plant biomass, but also enhance the plant absorption of heavy metals and other pollutants,or even induce the plant turn into hyperaccumulator, furthermore, improve the absolute and relative efficiency of plant extraction technology in heavy metal soil. Study has demonstrated that doubling CO_2 fertilizer under drought stress can significantly improve photosynthetic efficiency and antioxidant enzyme activities of wheat to increase heavy metal resistance. The soil test by Tang shown that[5] , elevating CO_2 concentration is not only beneficial to improve the ability of India mustard and sunflower to resist copper stress, but also significantly enhance their aboveground biomass, besides, induce the accumulation of copper in these two plants.

2.2. Use of other Agronomic Practices

The mainly agronomic practices that can be employed to maximize phytoremediation potential include tillage, co-cropping, crop rotation and coppicing[6-8]. Tillage is beneficial to turning heavy metals into the rooting zone,where remediation takes place, which promotes the contact between plants and heavy metals, and improves the efficiency of phytoremediation. According to the study[9], rotation, intercropping or interplanting are beneficial to reduce the influence of weeds, diseases and insect pests on plant growth and development. In the study, zinc, cadmium hyperaccumulator S.Alfredii and ordinary plants (maize) were intercropped in remediation of heavy metal contaminated soil.Meanwhile, there was a certain economic output, reducing the cost of soil remediation. Therefore, the application of intercropping or interplanting more than 2 hyperaccumulators can improve the remediation efficiency.

IOP Conf. Series: Earth and Environmental Science **78** (2017) 012015 doi:10.1088/1755-1315/78/1/012015

3. Earthwarm Strengthen Repair

Earthworms play an important role in agricultural production,it can also be used for phytoremediation, which is mainly reflected in two aspects: on the one hand, it can improve the physical and chemical properties of soil, develop soil fertility, and enhance plant biomass; on the other hand, earthworms may have direct or indirect effects on the bioavailability of heavy metals in soil while feeding, site preparation and excretion. Studies also found that earthworm activity in red soil can not only reduce pH value, but also significantly increase the extractable zinc content of DTPA, thus promoting the absorption and accumulation of plants[10].

4. Chemical Induction Technology

Most of the heavy metals in soil are difficult to be absorbed by plants. Chemical remediation agents,such as chelating agent, alkali substances, plant nutrients, ions material, plant hormones, humic acid, surface active agent, can increase the tolerance of plants to heavy metals and improve their ability to restore heavy metals[11]. This paper focuses on the chelating agent and pH regulator on Phytoremediation enhancement.

4.1. Organic Chelating Agent

The chelating agent can react with the heavy metal, which can be activated and improved its bioavailability. Further more ,the reaction can increase the uptake and transition of heavy metals by plants. Piechalak[12] reported that with the treatment of EDTA , in the soil content 200 mg/kg Pb,the accumulation of Pb in pea was increased by 67% compared with the control.

Despite the addition of chelating agent can maximize the effect of phytoremediation, there are potential environmental risks and disadvantages.The concentration of chelating agents is too high, which can cause toxicity to soil microorganisms and plants, inhibit the growth of plants, and cause heavy metal leaching and infiltration to groundwater, leading to groundwater pollution[13],[14].Therefore, the practicability and feasibility of chelating agents are still worth further study.

4.2. Acid-Base Regulator

The change of soil pH will directly affect the fraction and activity of heavy metals in soil. The use of pH regulator (elemental sulfur and calcium hydroxide), hyperaccumulator Thlaspicaerulescens at pH 5.84, the zinc and cadmium absorption reached the maximum with the increase of pH, Thlaspi enrichment of zinc and cadmium content decreased .Adding citric acid, malic acid and other low molecular weight organic acids in heavy metal contaminated soil, can promote the accumulation of heavy metal in plants, desorption in soil at the same time. And the degradation of organic acids in the soil as the final product are carbon dioxide and water, which are not prone to residue.

5. Other Strengthening Techniques and Measures

With the rapid development of molecular biology, genetic engineering technology, the metal chelating agent, metallothionein (MTs), phytochelatins (PCs) and heavy metal transporter genes can be transfered into hyperaccumulator, which will effectively increase the extraction plant of metal. Besides, the biomass of phytoremediation , tolerance and resistance to heavy metals can be improved simultaneously thanks to the transgenic technology[15]-[17].

In addition, inoculation of rhizosphere micro-organisms in heavy metal soil is also widely used. Soil microbial species, a large number of some rhizosphere microorganisms could change bioavailability of soil heavy metal through metals oxidation-reduction, or by the secretion of biosurfactant, organic acids, amino acids and enzymes to improve the biological availability of heavy metals in the root environment, promoting the growth of plants.

6. Perspectives and Suggestion

Though many achievements have been made in strengthening technology for phytoremediation of soil contaminated by heavy metals, in recent years, the study of these technologies needs further expansion. The following aspects should be carried out in the future:

1)The molecular mechanisms of the uptake, translocation, accumulation and detoxification of heavy metals in hyperaccumulator are still to be studied.

2)The specification of hyperaccumulators is having a strong enrichment effect of one kind heavy metal, which was limited to remediate contaminated soil which was usually caused by a variety of heavy metals.We should study on how to create plants that rich in heavy metals at the same time by means of genetic engineering.

3)Use of chelators and chemical surfactants, which have been shown to be toxic and can leach into ground water or soil, may have undesirable environmental consequences.Thus, it is necessary to develop a kind of natural or artificial chelating agent, which is low toxicity, easy degradation, high selectivity and high efficiency. It's better to combine with other strengthening techniques to maximize the efficiency of chelate induced.

4) Currently, systematic full-scale field studies should be conducted to explore a feasible plant cultivation management measures.Optimizing fertilization technology and applying new fertilizers to improve soil environment, promote plant growth and heavy metal absorption, and improve the efficiency of phytoremediation.

5)Study on the practice application of joint repair should strengthen various strengthening measures to jointly improve the phytoremediation of heavy metal contaminated soils.

6)The recovery process of low content heavy metals in plants still needs further study, so that the biomass of phytoremediation can be properly handled and environmental risks will be reduced.

7. References

[1] Marques A , Rangle A, and Castro P M 2011 *Crit. Rev. Environ. Sci. Technol.* 41 879-914

[2] Chaney R L, Malik M and Li Y M 1997 *Current Opinion in Biotechnology* 8 279-84

[3] Robinson B H 1997 The nickel hyperaccumulator plant Alyssum bertolon ii as a potential agent for phytoremediation and phytomining of nickel *Geochem.Explo.* 59 75-86

[4] Xiong X, Tang H, Huang S F 2012 Review in strengthening technology for phytoremediation of soil contaminated by heavy metals *Environ. Sci.Technol* 35 185- 93

[5] Tang S R, Xi L E and Zheng J M 2003 The responses of indian mustard and sunflower growing on copper contaminated soil to elevated CO_2 *Bulletin of Environ. Contami. Toxi.* 71 988-997

[6] Sheoran V, Sheoran A S and Poonia P 2016 Factors affecting phytoextraction: A review *Pedosphere* 26 148-66

[7] Lebeau T, Braud A and Jezequel K 2008 Performance of bioaugmentation-assisted phytoextraction applied to metal contaminated soils: a review *Environ Pollut* 153 497-522.

[8] Kaushal J, Bhasin S K and Bhardwaj P 2015 Phytoremediation: a review focusing on phytoremediation mechanisms *Int.J.Res.Chem.Environ.*5 1-9

[9] Sung K, Munster C L and Corapcioglu M Y 2004 Phytoremediation and modeling of contaminated soil using eastern gamagrass and annual ryegrass *Water, Air, and Soil Pollut* 159 175-95

[10] Nannoni F, Rossi S and Protano G 2014 Soil properties and metal accumulation by earthworms in the Siena urban area *Appli.Soi.Eco.*77 9-17

[11] Kumer A, Maiti S K, Tripti, Prasad M N V and Singh R S 2017 Grasses and legumes facilitate phytoremediation of metalliferous soils in the vicinity of an abandoned chromite-asbestos mine *J.Soi. Sedim.*17 1358-68

[12] Piechalak A, Tomaszewska B and kiewicz B D 2003 Enhancing phytoremediative ability of pisum sativum by EDTA application *Phytochemistry* 64 1239-51.

[13] Alvarez L V, Prieto F A, Cabello C M I and Kidd P S 2016 Organic amendments for improving biomass production and metal yield of Ni-hyperaccumulating plants *Sci.Environ* 54 370-9

[14] Kidd P, Mench M, Alvarez L V, Bert V, Dimitriou I, Friesl H W, Herzig R, Janssen J O, Kolbas A and Muller I 2015 *Internat. J.Phytorem.* 17 1005-37

[15] Marcela R, Maria A and Jose B 2017 Improvement of copper stress tolerance in pepper plants (Capsicum annuum L.) by inoculation with arbuscular mycorrhizal fungi *Theor.Exp.Plant Physiol.*29 37-49

[16] Eapen S and Souza S F 2005 Prospects of genetic engineering of plants for phytoremediation of toxic metals *Biotech. Adv* 23 97-114

[17] Chang P 2014 Plant growth-promoting bacteria facilitate the growth of barley and oats in salt-impacted soil: implications for phytoremediation of saline soils *Int.J.Phytorem.*16 1133-47

Acknowledgments

The project was supported by Guangxi Scientific Experiment Center of Mining, Metallurgy and Environment (KH2012ZD004), The Guangxi Talent Highland for Hazardous Waste Disposal Industrialization and The Special Funding for Guangxi 'BaGui scholars' Construction Projects.

MICP and Advances towards Eco-Friendly and Economical Applications

Adharsh Rajasekar[1], Charles K.S. Moy[2], Stephen Wilkinson[3]

1,2 Department of Civil Engineering, Xi'an Jiaotong-Liverpool University, Suzhou 215123, Jiangsu, China.
3 Department of Civil Engineering, University of Wolverhampton, WV1 1LY, United Kingdom.
Email: adharsh.rajasekar@xjtlu.edu.cn

Abstract. Biomineralization is a natural process aided by living organisms. Due to its applicability in ground improvement and bioremediation, Microbially Induced Calcite Precipitation (MICP) is an interdisciplinary field of study combining engineering, chemistry and microbiology. Bioremediation has been applied widely for contamination containment or removal, in this case it will be containment. MICP can also be applied to improve the efficiency of *insitu* bioremediation. Urease is an enzyme which can facilitate increased calcite precipitation. However the production of urease by bacteria and thus the resulting carbonate precipitation are inhibited by environmental factors including calcium concentration, bacterial concentration, pH and temperature. Under good conditions MICP can be used for heavy metal and radionuclide immobilization. However technologies such as bioconsolidation and biocementation require improvement such as time and cost. This paper highlights the application of MICP in addition to suggested improvements to make it more eco-friendly and sustainable.

1. Introduction

Microbial activity that alters the chemical environment favoring mineral formation is known as Biomineralization [1], [2]. It is a natural phenomenon that leads to precipitation of more than 60 different biological minerals that are formed through extracellular or intracellular pathways [3] . This occurs by a sequence of chemical reactions and physiological pathways which results in the precipitation of a range of different forms of solid mineral structure.

2. Microbially Induced Carbonate Precipitation

Microbially induced carbonate precipitation (MICP) is a biochemical mechanism which is driven by microorganism upon interacting with a chemical solution rich in calcium. Research in MICP has shown that microbially released CO_2 interacts with the biomineralization solution favoring carbonate formation. The carbonate combines with the calcium ion (Ca^{2+}) leading to the precipitation of calcium carbonate. In the last two decades, multiple mechanisms have been identified for the precipitation of calcium carbonate. The mechanisms include photosynthesis [4], [5], urea hydrolysis [1], [6]-[9], sulfate reduction [10], [11] and extracellular polymeric substances [12], [13]. Each mechanism promotes a different chemical pathway (Figure 1), all of which may be effective for mineralization. However, the precipitation of calcium carbonate by bacteria via urea hydrolysis is the most commonly exploited mechanism.

Urease activity is found in a wide range of microorganisms, one of the most commonly studied bacteria is *Sporosarcina pasteurii.* It is a soil, non-pathogenic and endospore producing, bacteria with an optimum pH for growth of 9.0 that can tolerate extreme conditions. Multiple studies have been conducted with *Sporosarcina pasteurii* for MICP [6], [8], [14]-[16]. Additionally, Achal, Mukherjee [17] developed a mutant strain (BP-M-3) of *Sporosarcina pasteurii* MTCC 1761 which resulted in an enhanced level of urease activity and carbonate precipitation compared to the natural type. The most important criteria to consider for the selection of a bacterial strain for biomineralization is its ability to synthesize active urease. However, a further consideration is that there are many pathogens among urease producing bacteria. For example, active urease producers includes *Helicobacter pylori* which infects the human stomach, and the opportunistic human pathogens such as *Proteus vulgaris*, *Staphylococcus aureus*, and *Pseudomonas aeruginosa* [18].

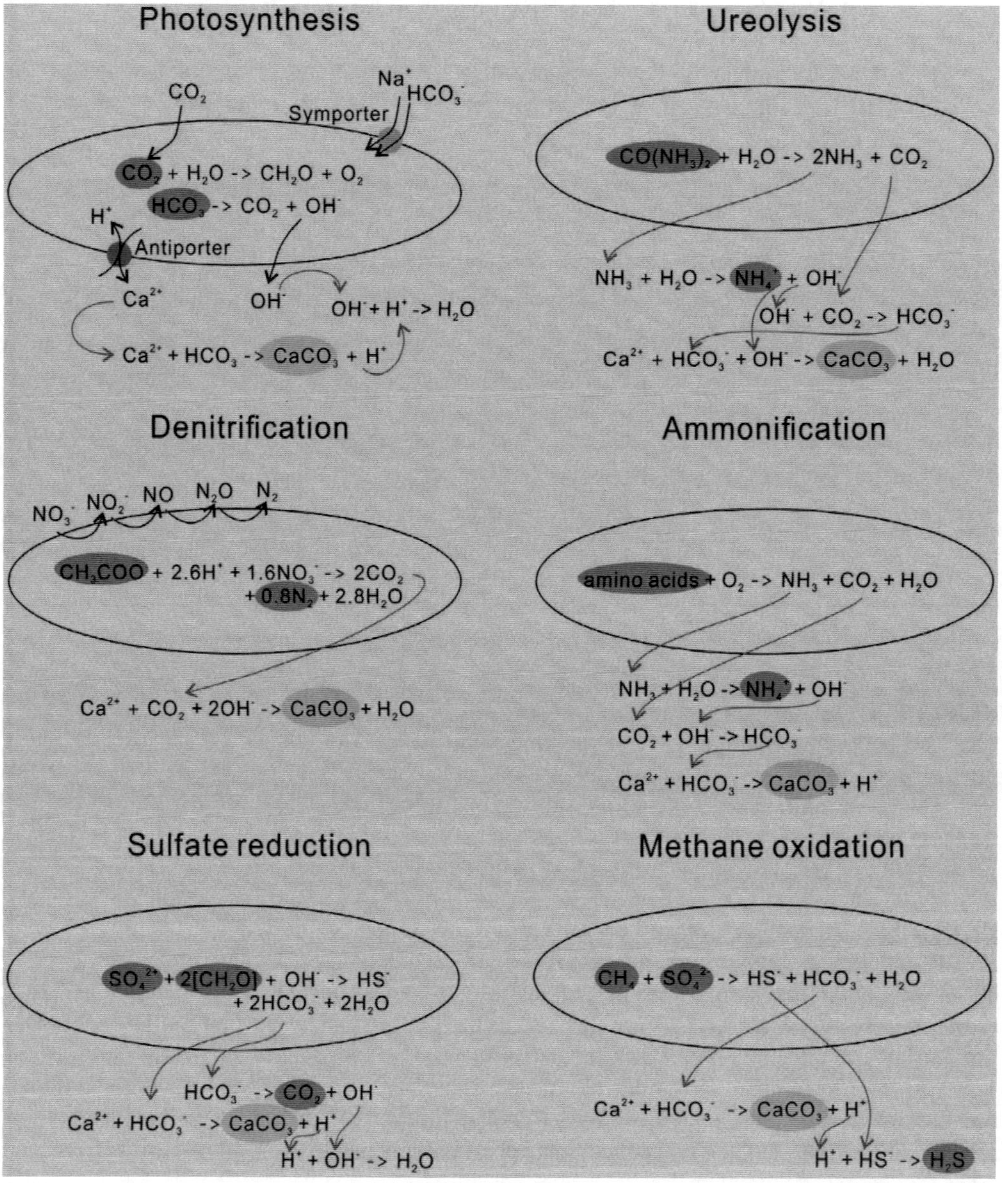

Figure 1. Processes that generate supersaturated environments essential for carbonate precipitation modified from [23].

Hammes and Verstraete [19] and Silva-Castro, Uad [20] reported that urease influences the chemical process associated with the formation of biominerals through four different factors; pH, dissolved inorganic carbon (DIC) concentrations, calcium concentrations and the availability of nucleation sites. The first three factors influence the carbonate ion concentration (CO_3^{2-}) while the last parameter promotes stable and continuous calcium carbonate formation [1], [21]. During the biomineralization process, bacteria commonly serve as nucleation sites for the precipitation of calcium carbonate. These four factors have a major influence on both ureolytic activity and calcium carbonate formation. Ca^{2+} ions bind to the negatively charged bacteria surfaces, creating a favorable environment for Ca^{2+} adsorption. Thus, Ca^{2+} ions bind more frequently onto the negatively charged cell surface of bacteria [22]. Bacterial cells are very important for the precipitation of $CaCO_3$, because the bacteria both provide nucleation sites (heterogeneous nucleation) and affect the types of minerals being formed (Figure 2). Okwadha and Li [8] found that a high concentration of bacterial cells increases the amount of carbonate precipitation via MICP. This occurs because of the increase in the concentration of urease increasing the rate of urea hydrolysis.

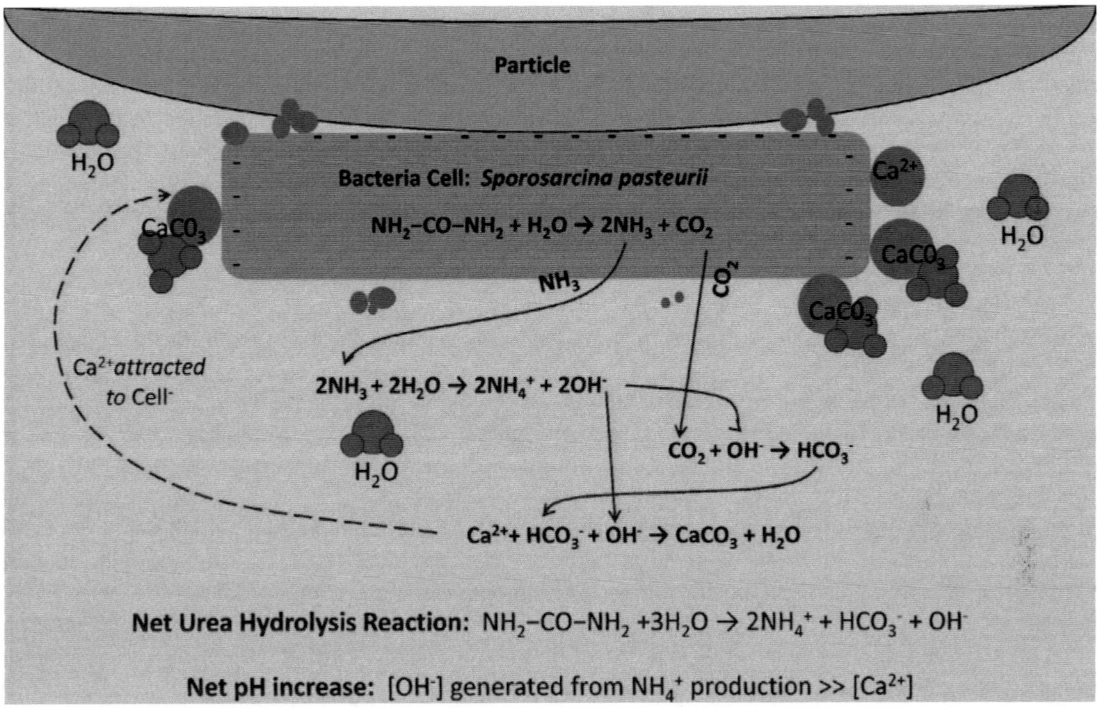

Figure 2. Schematic representation of Ureolysis in solution favoring MICP modified from [24].

3. Application towards Bioremediation

There have been a wide range of developments in bioremediation in recent years and these developments have their associated challenges which must be addressed for this technology to reach maturity as an engineering tool.

3.1. Removal of Heavy Metals and Radionuclides

Given the current rate of urbanization and industrialization, heavy metals and radioactive waste released both into the atmosphere are into soils due to industrial processes have been observed accumulating in both in landfills and residential environments [25]-[27]. These accumulated heavy metals and radionuclides pose serious health problems for humans and other living organisms within the environment. Some heavy metals in small dosages are beneficial to humans, but the rate of industrial release can be very toxic to humans [28]. The mobility of the released heavy metal ions may increase the seriousness of the threat to the lives of humans and effective methods need to be implemented to impede their transportation especially through groundwater [29].

Heavy metals including arsenic, cadmium and lead are commonly identified in most landfills at medium to high concentrations [26], [30]. Fu and Wang [28] proposed that heavy metals can be immobilized from the environment using MICP. However, heavy metal toxicity will also affect microbial growth and thus efficiency of MICP may be reduced; several researchers have identified and isolated heavy metal tolerant microbes with ureolytic capability from diverse environments which could improve the efficiency of the MICP process in contaminated ground [31], [32]. During the MICP process, calcium ions are added to a solution to precipitate calcium carbonate, in the heavy metal containment MICP process, calcium carbonates can also incorporate heavy metals (e.g., Cd and Pb^{2+}) into their surfaces via substitution of suitable divalent cations (Ca^{2+}) in the carbonate lattice, which alters the chemical form of these carbonates and alters the heavy metals from soluble to insoluble forms reducing their potential for toxicity.

3.2. Removal of Radionuclides

The disposal of radionuclide wastewater from commercial nuclear plants is a major issues associated with nuclear waste management because it is highly toxic to the environment, particularly to human health. Fujita, Taylor [9] assessed a pump and treat method, but it was unsuccessful at radionuclides removal from the contaminated environment. In such scenarios, MICP can be applied to immobilize the radionuclides safely from the environment. The basic process behind MICP method involves ureolytic microorganisms to precipitate $CaCO_3$, this in turn leads to promote co-precipitation of radionuclides by substitution of Ca^{2+} ion and formation of radionuclide carbonate minerals [9], [33].

3.3. Improvements Needed to Make MICP more Economical and Eco-Friendly

MICP has a great potential for sustainable environmental remediation. However as MICP is still a new methodology in terms of engineering application, there are a few limitations which must be addressed prior to field implementation:

1) MICP is not 100% environmentally friendly, as ureolysis plays a major role in precipitation generating by-products including ammonium and nitrate. These compounds are toxic and thus hazardous both to human health and to indigenous microbial consortia especially at high concentrations [34]. This limits its application for biocementation as ammonium present inside building materials have the potential to be converted into nitric acid by bacteria, which might decrease the bio-deterioration of materials. Ganendra, Muynck [35] found that replacing calcium chloride with calcium formate did not result in the release the ammonia to the air or produce nitric acid. More investigation and optimization is required to advance the process such that the volume/concentration of unwanted byproducts is reduced. Reduction of these byproducts would greatly improve the validity of the assessment that MICP is an eco-friendly treatment.

2) MICP is a microbial process which greatly depends on temperature, pH, calcium concentration, DIC and the presence of nucleation sites [36]. This makes it a complex and time consuming process in comparison to the chemical process under standard environmental conditions. MICP has to be optimized for time effectiveness before it's used for *insitu* applications.

3) The economic limitations makes MICP less industrially friendly, as laboratory grade sources needs to be used. Since there's a potential of inefficient MICP when using non-laboratory grade chemical reagents. Although alternative inexpensive nutrient sources for MICP such as lactose mother liquor have been implemented, consideration of a wider range of alternative sources would provide a better assessment of its cost effectiveness [17]. In addition to this limitation, application of *insitu* MICP would require the generation of substantial volumes of chemical reagents and microbial solutions. Although recently indigenous bacteria capable of MICP are reported, more studies that are target specific need to be implemented to resolve this issue [9], [37], [38].

Given the discussion above, although studies of MICP have generated promising results, its application at the large scale is still challenging. This technology is however worth of further study, and the resolution of the issues outlined would promote its implementation as a replacement for less sustainable alternative methods.

4. References

[1] Stocks-Fischer, S., J.K. Galinat, and S.S. Bang, *Microbiological precipitation of CaCO3.* Soil Biology and Biochemistry, 1999. 31: p. 1563-1571.

[2] Bazylinski, D.A., et al., *Controlled Biomineralization of Magnetite (Fe3O4) and Greigite (Fe3S4) in a Magnetotactic Bacterium.* Applied and Environmental Microbiology, 1995. 61(9): p. 3232–3239.

[3] Dhami, N.K., M.S. Reddy, and A. Mukherjee, *Biomineralization of calcium carbonates and their engineered applications: a review.* Frontiers in Microbiology, 2013. 4(314): p. 1-13.

[4] Bundeleva, I.A., et al., *Calcium carbonate precipitation by anoxygenic phototrophic bacteria.* Chemical Geology, 2012. 291: p. 116-131.

[5] McConnaughey, T.A. and J.F. Whelan, *Calcification generates protons for nutrient and bicarbonate uptake.* Earth-Science reviews, 1997. 42: p. 95-117.

[6] Cheng, L. and R. Cord-Ruwisch, *Selective enrichment and production of highly urease active bacteria by non-sterile (open) chemostat culture.* Journal of Industrial Microbiology and Biotechnology, 2013. 40(10): p. 1095-1104.

[7] Millo, C., et al., *Carbon isotope fractionation during calcium carbonate precipitation induced by urease-catalysed hydrolysis of urea.* Chemical Geology, 2012. 330-331: p. 39-50.

[8] Okwadha, G.D. and J. Li, *Optimum conditions for microbial carbonate precipitation.* Chemosphere, 2010. 81(9): p. 1143-1148.

[9] Fujita, Y., et al., *Stimulation Of Microbial Urea Hydrolysis In Groundwater To Enhance Calcite Precipitation.* Environmental science and technology, 2008. 42: p. 3025-3032.

[10] Deng, S., et al., *Microbial dolomite precipitation using sulfate reducing and halophilic bacteria: Results from Qinghai Lake, Tibetan Plateau, NW China.* Chemical Geology, 2010. 278(3-4): p. 151-159.

[11] Warthmann, R., et al., *Bacterially induced dolomite precipitation in anoxic culture experiments.* Geology, 2000. 28(12): p. 1091-1094.

[12] Tourney, J. and B.T. Ngwenya, *Bacterial extracellular polymeric substances (EPS) mediate CaCO3 morphology and polymorphism.* Chemical Geology, 2009. 262(3-4): p. 138-146.

[13] Li, W., et al., *Influence of initial calcium ion concentration on the precipitation and crystal morphology of calcium carbonate induced by bacterial carbonic anhydrase.* Chemical Engineering Journal, 2013. 218: p. 65-72.

[14] Gorospe, C.M., et al., *Effects of Different Calcium Salts on Calcium Carbonate Crystal Formation by Sporosarcina pasteurii KCTC 3558.* Biotechnology and Bioprocess Engineering, 2013. 18: p. 903-908.

[15] Tobler, D.J., et al., *Comparison of rates of ureolysis between Sporosarcina pasteurii and an indigenous groundwater community under conditions required to precipitate large volumes of calcite.* Geochimica et Cosmochimica Acta, 2011. 75(11): p. 3290-3301.

[16] Kang, C.H., et al., *Microbially induced calcite precipitation-based sequestration of strontium by Sporosarcina pasteurii WJ-2.* Applied Biochemistry and Biotechnology, 2014. 174(7): p. 2482-2491.

[17] Achal, V., et al., *Lactose mother liquor as an alternative nutrient source for microbial concrete production by Sporosarcina pasteurii.* Journal of Industraial Microbiology Biotechnology, 2009. 36(3): p. 433-438.

[18] Stabnikov, V., et al., *Halotolerant, alkaliphilic urease-producing bacteria from different climate zones and their application for biocementation of sand.* World Journal of Microbiology and Biotechnology, 2013. 29: p. 1453-1460.

[19] Hammes, F. and W. Verstraete, *Key roles of pH and calcium metabolism in microbial carbonate precipitation.* Re/Views in Environmental Science & BioTechnology, 2002. 1: p. 3-7.

[20] Silva-Castro, G.A., et al., *Carbonate Precipitation of Bacterial Strains Isolated from Sediments and Seawater: Formation Mechanisms.* Geomicrobiology Journal, 2013. 30(9): p. 840-850.

[21] Lian, B., et al., *Carbonate biomineralization induced by soil bacterium Bacillus megaterium.* Geochimica et Cosmochimica Acta, 2006. 70(22): p. 5522-5535.

[22] Sanchez-Roman, M., et al., *Biomineralization of carbonate and phosphate by moderately halophilic bacteria.* FEMS Microbiology Ecology, 2007. 61(2): p. 273-284.
[23] Zhu, T. and M. Dittrich, *Carbonate Precipitation through Microbial Activities in Natural Environment, and Their Potential in Biotechnology: A Review.* Frontiers in Microbiology, 2016. 4(4): p. 1-21.
[24] DeJong, J.T., et al., *Bio-mediated soil improvement.* Ecological Engineering, 2010. 36: p. 197-210.
[25] Zhang, D.Q., S.K. Tan, and R.M. Gersberg, *Municipal solid waste management in China: status, problems and challenges.* Journal of Environmental Management, 2010. 91(8): p. 1623-1633.
[26] Pablos, M.V., et al., *Correlation between physicochemical and ecotoxicological approaches to estimate landfill leachates toxicity.* Waste Management, 2011. 31(8): p. 1841-1847.
[27] Olivero-Verbel, J., C. Padilla-Bottet, and O. De la Rosa, *Relationships between physicochemical parameters and the toxicity of leachates from a municipal solid waste landfill.* Ecotoxicology and Environmental Safety, 2008. 70(2): p. 294-299.
[28] Fu, F. and Q. Wang, *Removal of heavy metal ions from wastewaters: A review.* Journal of Environmental Management, 2011. 92: p. 407-418.
[29] Mor, S., et al., *Leachate characterization and assessment of groundwater pollution near municipal solid waste landfill site.* Environmental Monitoring and Assessment, 2006. 118(1-3): p. 435-456.
[30] Perez-Leblic, M.I., et al., *Influence of xenobiotic contaminants on landfill soil microbial activity and diversity.* Journal of Environmental Management, 2012. 95: p. 285-290.
[31] Kang, C.-H., et al., *Bioremediation of Cd by Microbially Induced Calcite Precipitation.* Applied Biochemistry and Biotechnology, 2014. 172: p. 2907-2915.
[32] Li, M., X. Cheng, and H. Guo, *Heavy metal removal by biomineralization of urease producing bacteria isolated from soil.* International Biodeterioration & Biodegradation, 2013. 76: p. 81-85.
[33] Mitchell, A.C. and F.G. Ferris, *The coprecipitation of Sr into calcite precipitates induced by bacterial ureolysis in artificial groundwater: Temperature and kinetic dependence.* Geochimica et Cosmochimica Acta, 2005. 69(17): p. 4199-4210.
[34] Paassen, L.A.v., et al., *Quantifying Biomediated Ground Improvement by Ureolysis: Large-Scale Biogrout Experiment.* Journal of Geotechnical and Geoenvironmental Engineering, 2010. 136: p. 1721-1728.
[35] Ganendra, G., et al., *Formate Oxidation-Driven Calcium Carbonate Precipitation by Methylocystis parvus OBBP.* Applied and Environmental Microbiology, 2014. 80(15): p. 4659–4667.
[36] Ivanov, V. and J. Chu, *Applications of microorganisms to geotechnical engineering for bioclogging and biocementation of soil in situ.* Reviews in Environmental Science and Bio/Technology, 2008. 7(2): p. 139-153.
[37] Zamarreño, D.V., E. May, and R. Inkpen, *Influence of Environmental Temperature on Biocalcification by Non-sporing Freshwater Bacteria.* Geomicrobiology Journal, 2009. 26(4): p. 298-309.
[38] Yoshida, N., E. Higashimura, and Y. Saeki, *Catalytic Biomineralization of Fluorescent Calcite by the Thermophilic Bacterium Geobacillus thermoglucosidasius.* Applied and Environmental Microbiology, 2010. 76(21): p. 7322-7327.

Acknowledgments
This work was supported by grant no. PGRS-12-02-06 and RDF-13-01-06 awarded by XJTLU.

Isolation and Characterization of Biosurfactant Producing Bacteria for the Application in Enhanced Oil Recovery

Niraj Prasad[1], Sumita Dasgupta[1], Mousumi Chakraborty[2], Smita Gupta[2,*]

1 Bhagwan Mahavir College of M. Sc. Biotechnology, Surat - 395 007, Gujarat, India
2 Department of Chemical Engineering, Sardar Vallabhbhai National Institute of Technology, Surat - 395 007, Gujarat, India
*Email: smitagupta12@rediffmail.com, g.smita@ched.svnit.ac.in

Abstract. In the present study, a biosurfactant producing bacterial strain was isolated, screened and identified. Further, various fermentation conditions (such as pH (5-10), incubation period (24-96h) and incubation temperature (20-60 °C) were optimized for maximum production of biosurfactant. The produced biosurfactant was characterized by measuring emulsification index, foaming characteristics, rhamnolipid detection, interfacial tension between water and oil and stability against pH and temperature for its potential application in oil recovery process. The additional oil recovery for two different sand, sand1 and sand2, was found to be 49% and 38%, respectively.

1. Introduction

The percentage of oil remaining in a reservoir varies from field to field. However, in a recent study, it was found that 65% of the original oil in place (OOIP) was left behind after primary and secondary-oil recovery processes [1]. In general residual oil can be recovered by three EOR technologies i.e. thermal, miscible and chemical enhanced oil recoveries (CEOR). Among these, CEOR is one of the most common methods of EOR. This method involves the recovery of residual oil by using different types of chemical formulations which are capable enough in reducing mobility ratio and/or increasing capillary number when injected in the reservoir as displacing fluid. Use of chemical surfactant for this purpose is one of the common method [2]. But, chemical surfactants are having many disadvantages in environmental as well as application point of view. They can be hazardous, non biodegradable, toxic to the environment, and costly compounds [3]. Moreover, with these compounds achieving the ultra low interfacial tension (i.e. less than 10^{-3} mN/m), which is one of the basic requirement for EOR, is one of the big challenges in the field of EOR. Because, the interfacial tension (IFT) between water and oil is directly related with the capillary forces and high capillary force resulted into the entrapment of the oil in the porous matrix of the reservoir [4].

From literature it was found that various articles are available on MEOR from 2008 onwards, using biosurfactants produced from various microorganisms (Table 1) but still there is a scope in achieving the desired characteristics of interfacial tension comparatively more nearer to ultralow level (i.e. 10^{-3} mN/m) and improving the MEOR by reducing the capillary forces for channelizing the flow of oil through the porous matrix of the reservoir. For this purpose, an effort has been made to replace chemical surfactant with the eco-friendly bio-surfactant, which is produced ex-situ from an isolated bacterial strain, characterized in terms of interfacial tension and used for the application in microbial enhanced oil recovery.

Table 1. Literature Review on MEOR

Microorganism for MEOR	Oil Recovery (%)	Reduction in Surface tension (ST)/IFT	Reference
Bacillus subtilis PT2 and *Pseudomonas aeruginosa* SP4	61.62 and 57.07, respectively	ST= 26.4 and 28.3mN/m, respectively	[5]
Bacillus amyloliquefaciens BZ-6	88% from oily sludge	-	[6]
Bacillus subtilis	-	ST=30 mN/m	[7]
Bacillus subtilis B20	9.7	ST=25 mN/m, IFT = 5.02 mN/m	[8]
Bacillus subtilis B30	light oil=26, heavy oil=31	ST=26 mN/m, IFT=4 mN/m	[9]
Pseudomonas sp. SWP- 4	24	ST=22.7 mN/m	[10]
Pseudomonas aeruginosa Pa4, *Escherichia coli*	9.96 and 12, respectively	IFT= 32.39 mN/m and 27.39 mN/m, respectively	[11]
Pseudomonas putida	26.1	IFT=2.5 mN/m	[12]

2. Experimental Procedure

Please follow these instructions as carefully as possible so all articles within a conference have the same style to the title page. This paragraph follows a section title so it should not be indented.

2.1. Materials and Methods

The sample collection was aimed for isolation of biosurfactant producing organism. For this purpose, the soil sample was collected near automobile garage in Udhana, Surat-Gujarat analyzed for isolation of oil degrading microorganism. The temperature of the soil was around 28 °C and pH 8.0. Bushnell Hass Mineral Salts (BHMS) medium (purchased from Himedia, Mumbai, India) was used for the isolation of hydrocarbon degrading bacteria. The pH was adjusted to 7. The BHMS medium was supplemented with 1% (v/v) used engine oil as the sole carbon source [7]. For MEOR experiment, paraffin liquid light oil (Finar Chemicals) was used asthe oil phase. Sodium chloride (99%) for salinity and Cetyl trimethylammo-nium bromide (CTAB) (99%) for rhamnolipid detection, were purchased from Sigma-Aldrich(India). All other chemicals like HCl and NaOH used were of analytical grade. All the solutions were prepared in demineralized water. Interfacial tension was estimated between oil (light paraffin oil) phase and water with different concentrations of the produced biosurfactant using a KRUSS-T9Tensiometer (Germany), (Du-Nuoy ring method) under atmospheric pressure.

2.2. *Experimental Setup*

All fermentation experiments were performed in 500 ml conical flasks. MEOR experiments were performed in 60 ml syringes, packed with two types of sand (calcite (sand 1) and silicate (sand 2)). Four sets were prepared for two different samples among them one served as control for each sand sample. For this test each syringe was packed with sterile sand 40 gram and was packed tightly using syringe plungers. The end nozzle was temporarily sealed. The experiment was carried out at 30°C.
The syringe system was flooded with water at a constant flow rate 1ml/min. Pore volume (PV, ml) was calculated using following equation:

$$\mathrm{PV} = \frac{\textbf{weight of core 100\% saturated with water} - \textbf{weight of dry core}}{\textbf{density of water}}$$

Porosity(%) of the system was measured by dividing the PV by the total volume of packed syringed with sand(20ml). In the Second step, the paraffin oil was injected into the system to replace water in all the syringe, until there was no water coming out from the nozzle in effluent. Original Oil in Place (OOIP) was calculated as the volume of paraffin oil retained in the system. Thus from this Initial Oil saturation (S_{oi}, %) and initial water saturation (S_{wi}, %) were calculated by following formula:

$$S_{oi} = \frac{OOIP}{PV} \times 100$$

$$S_{oi} = \frac{PV - OOIP}{PV} \times 100$$

Following step was to stabilize the system for 24 h and was flooded again to remove out the excess oil until no more paraffin oil was observed in the water effluent. Oil recovered after flooding (S_{orwf},ml) was determined using volumetric flask. Residual oil saturation (S_{or}, %) was calculated as follow:

$$S_{or} = \frac{OOIP - S_{orwf}}{OOIP} \times 100$$

Finally, the residual oil in the system was supplemented with the supernatant containing biosurfactant and for control only BHM broth was utilized. This system was maintained at 30°C for 48 h at shaker with 100rpm. After incubation the system was again flooded to recover the oil. The recovered oil was measured volumetrically and denoted as S_{orbf}(ml).
The Additional oil recovery (MEOR) was calculated as follow:

$$MEOR\ (\%) = \frac{S_{orbf}}{OOIP - S_{orwf}} \times 100$$

2.3. *Characterization and Identification of Isolate and Biosurfactant*

The isolated microorganisms were identified by standard gram staining, motility test, colony characterization of isolates, biochemical characterization of isolates and Molecular identification by 16s rRNA sequencing and the produced biosurfactant was characterized by performing emulsification index E24 test, Foaming test, rhamnolipid detection and oil displacement test [13].

3. Results and Discussion

3.1. *Isolation, Screening and Characterization of Biosurfactant Producing Bacteria*

All 14 isolates were screened with oil displacement test for biosurfactant production on used engine oil. Among these isolated 14 organisms, 2 showed oil displacement and formed a clear halo. Maximum oil displacement was seen in one of the isolates with halo diameter of 72 mm (Figure 1(a)) which confirmed its ability of producing biosurfactant and selected for further characterization, optimization and oil recovery experiments.

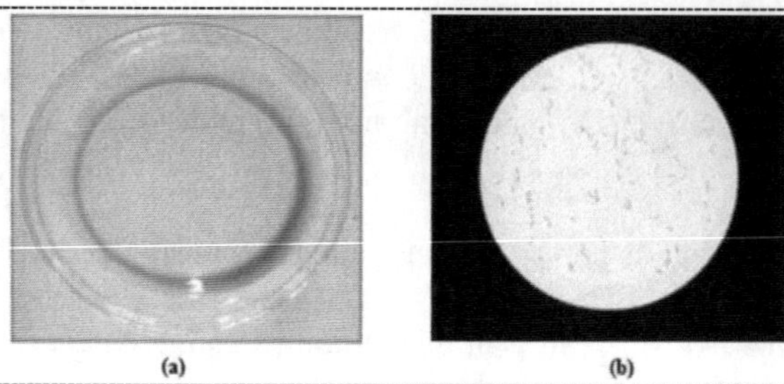

Figure 1 (a) Oil displacemnt zone of halo. **(b)** Microscopic view of Gram stained isolated strain of bacteria.

The isolated strain was found to be gram negative rod shaped and highly motile bacterial strain (Figure 1(b)). Further, molecular identification was done using 16s rRNA sequencing. The obtained sequence was search for its similarity using nucleotide BLAST 2.2.31 (blastn) software. The identification was made evaluating maximum score obtained, coverage, expected Value, and percentage identity. The isolate was confirmed to be Genus-*Pseudomonas* species *aeruginosa* with 99% similarity.

3.2. Effect of Incubation Period, pH and Incubation Temperature on Biomass and Biosurfactant Production

To produce the maximum amount of biosurfactant from the isolated bacteria during the fermentation process, various parameters of the fermentation process needs to be optimized. In this regard, various series of experiments for the fermentation of isolated strain were performed at different conditions of different process parameters like pH (varied from 5-10), temperature (varied from 20 °C to 50 °C) and time-period (varied from 24 to 96 h) for production of biosurfactant. For this purpose, the culture was inoculated with the inoculums size of 3.9×10^7 CFU/ml in BHM broth supplemented with 0.1M Glucose as a sole carbon source and the parameters were optimized by measuring biomass (dry weight) and oil displacement. The results of these experiments are shown in Figure 2. These results indicate that the biosurfactant synthesized from the isolated bacterial strain occurred predominantely during the exponential growth phase, suggesting that the biosurfactant was produced as primary metabolite accompanying cellular biomass formation (growth-associated kinetics) [14]. Based on these experiments, the optimum parameters for the maximum production of biosurfactant were attained by incubating 3.9×10^7 CFU/ml of the inoculum, upto the incubation period of 96 h at 7.0 pH and 37 °C.

3.3. Characterization of Crude Biosurfactant for Its Application in Enhanced Oil Recovery

In order to check the applicability of the produced biosurfactant, it is necessary to characterize it in terms of emulsification index (E24), foaming, rhamnolipid detection and interfacial tension at different conditions (pH, salinity, high temperature and pressure). The method of determining these characteristics are described elsewhere [13]. The results of these experiments are shown in Figure 3. The interfacial tension was measured against paraffin oil. A significant reduction (98% with respect to water) in the interfacial tension (nearer to ultralow) between water and paraffin was observed using the produced biosurfactant, which is desirable for the application enhanced oil recovery. Recently, minimum reduction in interfacial tension using conventional surfactant (CTAB) and ionic liquid surfactant (C_{16}mimBr) achieved were 7.7 mN/m and 3.8 mN/m, respectively [15].

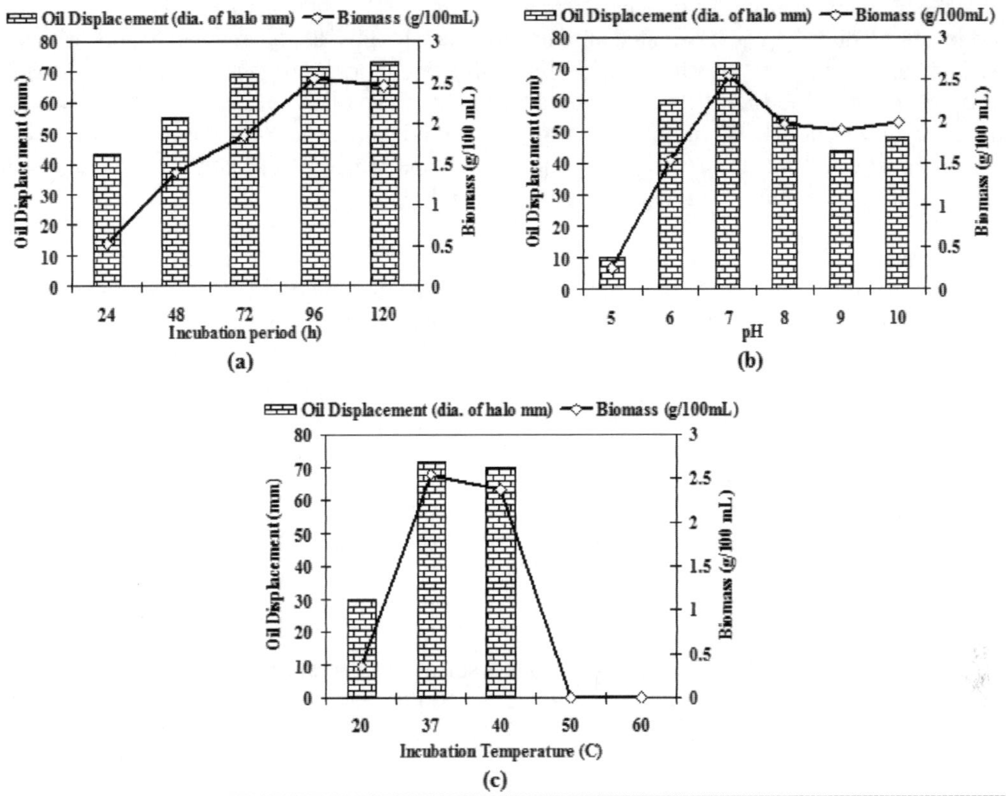

Figure 2. Effect of (a) incubation period with pH 7 and at 37 °C of incubation temperature, (b) pH at 96 h of incubation period and 37 °C of incubation temperature and (c) temperature at pH of 7 and 96 h of incubation period, fermentation and biosurfactant production.

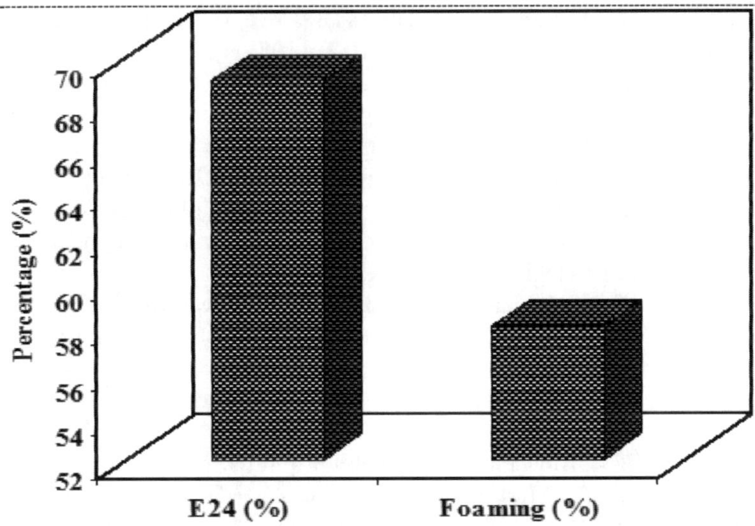

Figure 3. Emulsification index E24 and Foaming detection of the biosurfactant produced from the isolated strain.

3.4. MEOR Column Assay
Biosurfactant produced by the isolated strain was used in this assay and have high potential in decreasing surface tension in varied pH, high temperature and pressure. It shows a significant reduction of interfacial tension between water and light paraffin oil. Thus, it shows good potential for its utilization in MEOR. The additional oil recovery for sand1 and sand2 was measured after nullifying the control volume and was found to be 49% and 38%, respectively. This shows that the biosurfactant produced from isolated strain is highly efficient for MEOR.

4. Conclusions
The isolated strain was biosurfactant producing organism and indentified as Genus-*Pseudomonas* species *aeruginosa.* The isolate showed optimum growth at pH 7.0 and temperature 37°C. The biosurfactant produced by the isolate was detected as rhamnolipid and having significant reduced interfacial tension with good stability at harsh condition of the reservoir. The results show that the produced biosurfactant is highly efficient for MEOR.

5. References
[1] V. Pillai, J.R. Kanicky, D.O. Shah, Application of microemulsion in enhanced oilrecovery, in: P. Kumar, K.L. Mittal (Eds.), Handbook of MicroemulsionScienceand Technology, Marcel Dekker Inc., New York, 1999, pp. 743–754.

[2] V. Alvarado, E. Manrique, Enhanced oil recovery: an update review, Energies 3(2010) 1529–1575.

[3] N.K. Bordoloi & B.K. Konwar, Microbial surfactant-enhanced mineral oil recovery under laboratory conditions, Colloids and Surfaces B: Biointerfaces, 63(2008)73–82I.

[4] E.J. Gudina, J.F. Pereira, R. Costa, J.A. Coutinho, J.A. Teixeira, L.R. Rodrigues, Biosurfactant-producing and oil-degrading Bacillus subtilis strains enhance oil recovery in laboratory sand-pack columns, J Hazard,261(2013)106–113.

[5] O. Pornsunthorntawee, N. Arttaweeporn, S. Paisanjit, P. Somboonthanate, M. Abe, R. Rujiravanit, S. Chavadej, Isolation and comparison of biosurfactants produced by *Bacillus subtilis* PT2 and *Pseudomonas aeruginosa* SP4 for microbial surfactant-enhanced oil recovery, Biochemical Engineering Journal 42 (2008) 172–179.

[6] W. Liu, X. Wang, L. Wu, M. Chen, C. Tu, Y. Luo, P. Christie, Isolation, identification and characterization of Bacillus amyloliquefaciens BZ-6, a bacterial isolate for enhancing oil recovery from oily sludge, Chemosphere 87 (2012) 1105–1110.

[7] E.J. Gudina, J.F.B. Pereira, L.R. Rodrigues, J.A.P. Coutinho, J.A. Teixeira, Isolation and study of microorganisms from oil samples for application in Microbial Enhanced Oil Recovery, International Biodeterioration & Biodegradation 68 (2012) 56-64.

[8] S.N. Al-Bahry, Y.M. Al-Wahaibi, A.E. Elshafie, A.S. Al-Bemani, S.J. Joshi,H.S. Al-Makhmari, H.S. Al-Sulaimani, Biosurfactant production by Bacillus subtilis B20 using date molasses and its possible application in enhanced oil recovery, International Biodeterioration & Biodegradation 81 (2013) 141-146.

[9] Y. Al-Wahaibi, S. Joshi, S.Al-Bahry, A. Elshafie, A. Al-Bemani, B. Shibulal, Biosurfactant production by Bacillus subtilis B30 and its application inenhancing oil recovery, Colloids and Surfaces B: Biointerfaces,114(2014)324–333.

[10] G. Lan, Q. Fan, Y. Liu, Y. Liu, Y. Liu, X. Yin, M. Luo, Effects of the addition of waste cooking oil on heavy crude oil biodegradation and microbial enhanced oil recovery using Pseudomonas sp. SWP- 4, Biochemical Engineering Journal,103(2015)219–226.

[11] A.C. Alvarez Yela, M.A.T. Martínez, G.A.R. Pineros, V.C. Lopez, S.H. Villamizar, V.L.N. Velez, W.-R. Abraham, M.J.V. Florez, A.F.G.Barrios, A comparison between conventional Pseudomonas aeruginosa rhamnolipids and Escherichia coli transmembrane proteins for oil recovery enhancing, International Biodeterioration & Biodegradation 112 (2016) 59-65

[12] P. Sivasankar, G. Suresh Kumar, Influence of pH on dynamics of microbial enhanced oil recovery processes using biosurfactant producing Pseudomonas putida: Mathematical modelling and numerical simulation, Bioresource Technology, 224(2017)498–508.

[13] V.M. Alvarez, D. Jurelevicius, J.M. Marques, P. Macedo de Souzaa, L. Vieira de Araujo, T. G. Barros, R.O.M. Alves de Souza, D.M.G. Freire, L. Seldin, Bacillus amyloliquefaciens TSBSO 3.8, a biosurfactant-producing strainwith biotechnological potential for microbial enhanced oil recovery, Colloids and Surfaces B: Biointerfaces,136 (2015) 14–21.
[14] A. Persson, E. Österberg, M. Dostalek, Biosurfactant production by Pseudomonas fluorescens 378: growth and product characteristics, Applied Microbiology and Biotechnology 1988, 29(1), 1–4.
[15] S.K. Nandwani, N.I. Malek, V.N. Lad, M. Chakraborty, S. Gupta, Study on interfacial properties of Imidazolium ionic liquids as surfactant and their application in enhanced oil recovery, Colloids and Surfaces A: Physicochemical and Engineering Aspects, Colloids and Surfaces A: Physicochem. Eng. Aspects, 516 (2017) 383–393.

Optimal Energy Management of an Academic Building with Distributed Generation and Energy Storage Systems

C Roldán-Blay[1,2], C Roldán-Porta[1], E Peñalvo-López[1] and G Escrivá-Escrivá[1]

1 Institute for Energy Engineering, Universitat Politècnica de València, Camino de Vera, s/n, edificio 8E, escalera F, 5ª planta. 46022 Valencia, Spain.
2 Corresponding author. E-mail: carrolbl@die.upv.es.

Abstract. In this paper, an optimisation algorithm is used to simulate the management of distributed energy resources in an academic building. This optimisation algorithm, called DEROP, consists of an iterative procedure reach a supply schedule with the minimum energy cost. The inputs to the algorithm are the demand forecast, the availability of each resource, the level of storage in energy storage systems and prices and efficiencies of each resource. With these data, the algorithm proposes the optimal schedule to minimise costs of energy supply. The main advantages of this algorithm are that it is fast, easy to be implemented in real buildings and flexible. The algorithm is simulated with real data to optimise management of distributed energy resources and energy storage systems in an academic building. The management of these resources is optimised for a tariff with hourly discrimination and for a tariff with no time restrictions. One of the main conclusions drawn from these simulations are that significant savings are obtained with this algorithm. Also, DEROP allows taking advantage of tariffs with hourly discrimination, even in an academic building with low night-time consumption in which, a priori, these tariffs are not profitable.

1. Introduction

In order to minimise the dependence on fossil fuels, new energy resources are being integrated in energy systems [1], such as renewable energy sources (RESs) – wind, solar, biomass and so on – and energy storage systems (ESSs). With the increase of RESs, optimal management and control of all the available resources is a key issue to be addressed and many research studies have been developed over the past years [2]. This is usually studied from the perspective of energy generation facilities [3]. Nonetheless, a perspective of final customers with distributed generation (DG) and ESSs [4] might provide better optimisation results. This perspective is focused on facilities with multiple energy sources, ESSs and loads, which are widely called energy hubs.

From the perspective of end users, optimal management of their available resources consists of controlling all energy flows in their facilities (between power grid, DG resources, ESSs and loads) to minimise the total energy costs. This requires robust energy control systems (with real-time data acquisition and processing from loads, generation resources availability, external variables such as temperature or wind speed, energy purchase prices, forecasts...) and sophisticated algorithms to achieve an optimal management of the available resources along the time.

Several algorithms to optimise energy hubs' operation have been proposed, such as in [5]. Some of them are focused on demand side management [6]. Others describe algorithms to achieve optimal management of DERs and loads along a day [7]. The main disadvantages of these algorithms are their complexity, the difficulty to implement them in a real facility and the fact that only linear functions may be used in most of them, as in [8]. Other works focus on sources management, such as [9].

However, these works develop algorithms mainly suitable for microgrids, such as [10]. In this paper, a new algorithm for DERs optimisation in energy hubs, called DEROP, is used to minimise energy costs in an academic building. The inputs to DEROP are forecasts of demand and generation of each resource, initial level of charge in ESSs, prices and efficiency curves of each resource. From these data, DEROP calculates the optimal schedule for the energy hub using an iterative procedure. The main advantages of DEROP algorithm are that it is simple, fast and easy to be implemented in real buildings. Moreover, it is flexible, as it allows the use of non-linear functions for prices and efficiencies and users may add as many resources as needed.

This paper is organised as follows. Section 2 describes DEROP algorithm and the scenarios to be simulated. Section 3 shows a simulation step by step. Section 4 shows of the simulated scenarios. These results are compared, analysed and discussed in this section. Finally, some conclusions to this work are drawn in section 5.

2. Methodology

As aforementioned, an energy hub is a building or a set of buildings with several generation resources, loads and ESSs. This concept is shown in Figure 1. The goal of DEROP algorithm is to find the optimal schedule to manage DG resources, ESSs and power grid in an energy hub, to achieve the minimum energy cost. In order to do this, DEROP consists of two stages. The first stage optimises the use of DG resources. It tries to maximise the usage of RESs, using ESSs to store surplus generation and discharging ESSs in the most expensive times to allow storing energy when needed. The second stage optimises the use of power grid and ESSs. To do this, ESSs are optimally managed to allow grid disconnections in the most expensive times and charge ESSs in the cheapest times. This process is illustrated in Figure 2.

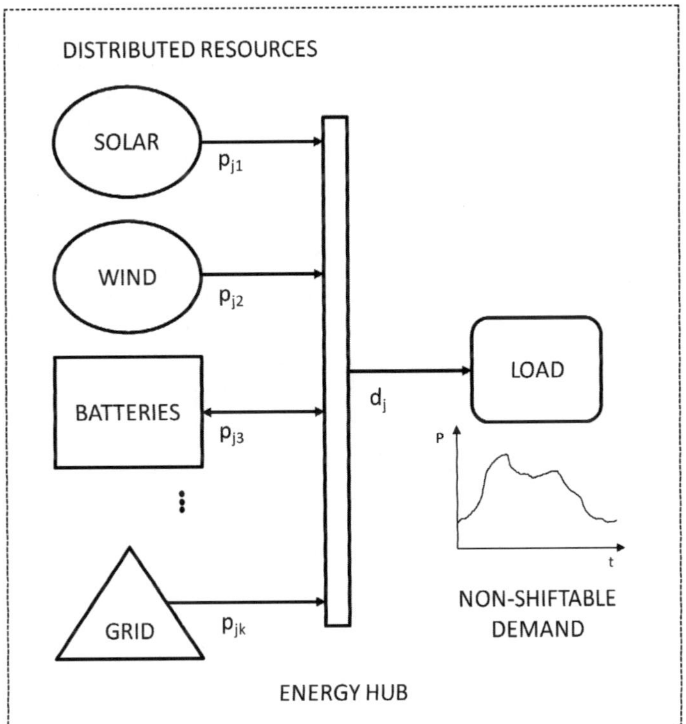

Figure 1. Energy hub concept to implement DEROP algorithm in order to optimise DERs management.

The algorithm has been tested in an academic building, where loads cannot be managed. Real data of generation and consumption curves have been used to simulate an optimal management with DEROP.

IOP Conf. Series: Earth and Environmental Science **78** (2017) 012018 doi:10.1088/1755-1315/78/1/012018

These data correspond to wind and solar generation curves measured in the Laboratory of Distributed Energy Resources (LabDER) at Universitat Politècnica de València (UPV) and demand curves of a departmental building at UPV (building 5E) that has laboratories, classrooms, seminars and offices. Also, real prices have been downloaded from the System Operator's website [11]. With these data, DEROP starts calculating total energy costs of an initial schedule (consisting of only power grid supply) and, through an iterative procedure, it creates new schedules that gradually decrease these costs until the last iteration, where no improvement can be made.

The simulated energy hub corresponds to an academic building with a maximum demand of over 100 kW, with solar panels and a wind generator (70 kW installed of each resource) and batteries (48 V and 4000 Ah to store nearly 200 kWh).

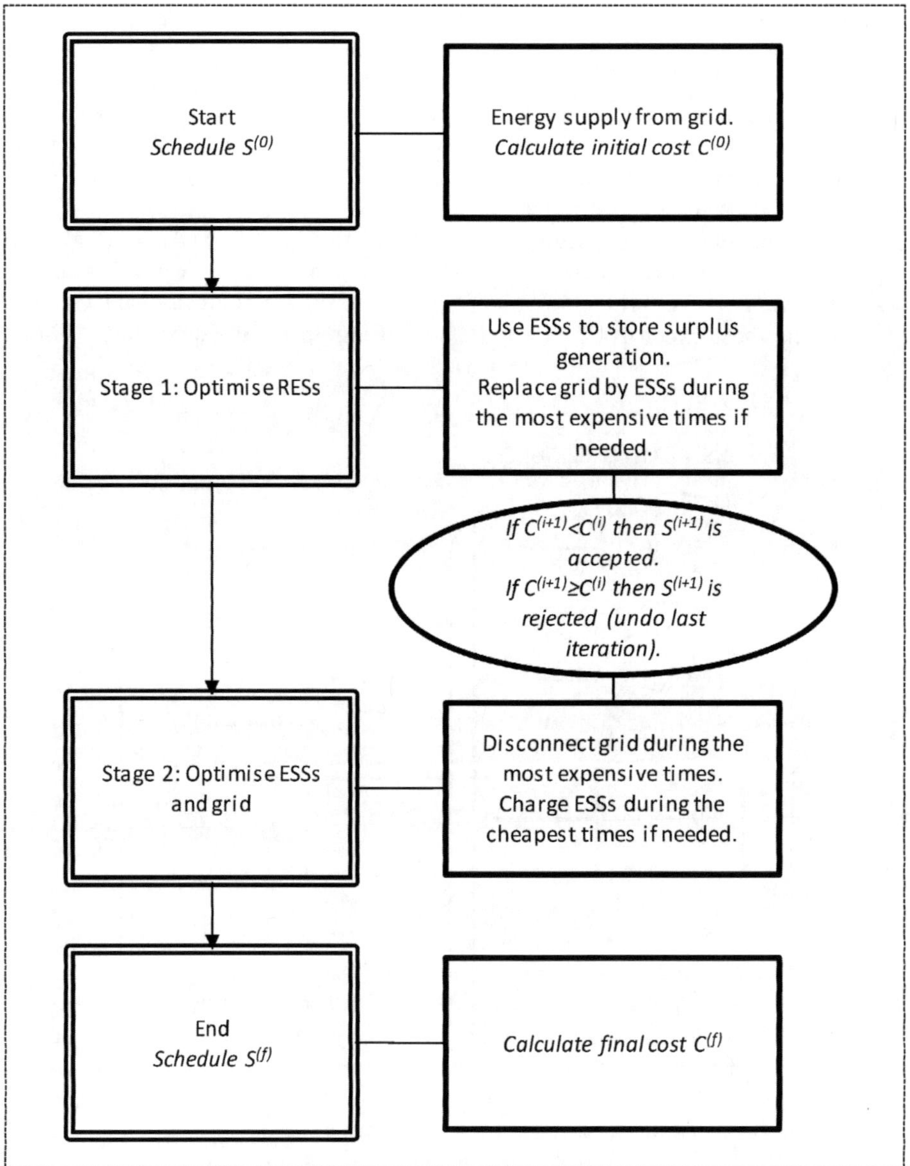

Figure 2. DEROP algorithm diagram.

A cost has been associated with each resource. In addition, each resource has its own efficiency curve. One of the main advantages of DEROP is its flexibility, since non-linear functions may be used for both, costs and efficiencies. Furthermore, DEROP allows users to add as many resources as needed to optimise their management.

To test the algorithm in different situations, the same curves have been used to optimise the management of resources with two different tariffs, i.e. a tariff with hourly discrimination (scenario A) and a tariff with no time restrictions (scenario B).

3. Scenarios Simulation

The described scenarios have been simulated using DEROP algorithm with real data. The simulation of scenario A is shown below.

In this scenario there are RESs (wind and solar power) and an ESS (batteries). This scenario has an electricity tariff with hourly discrimination. The goal of this simulation is to maximise the use of RESs and manage ESSs usage to minimise overall energy costs, using DEROP algorithm.

At the beginning of the simulation, in iteration i=0, the initial schedule $S^{(0)}$ consists of using grid supply for the entire simulation period (one day). Energy costs in this iteration are € 97.14, with a total demand of 1033.78 kWh. The associated emissions are 257.6 kg of CO_2.

In the 1st iteration the available RESs are used and surplus generation is used to charge ESSs. The result of this step is shown in Figure 3. The total cost of supplied energy in this case is € 78.02 and the total emissions are 186.3 kg of CO_2.

To complete the 2nd iteration DEROP algorithm looks for interval the $[t_j,t_{j+1}]$ with the highest purchase price (which takes place during the interval 19:00-19:15) and grid supply is disconnected during that time interval. To perform this simulation the selected step size is 15 minutes, although the smaller step sizes, the more accurate results may be obtained (for example, a step size of one minute could be used). As a result of this iteration the energy cost has decreased until € 75.28 and total emissions are 181.3 kg of CO_2. In the next iterations new grid disconnections are scheduled in expensive intervals, until a moment when battery is not able to meet demand. In iteration i=6, if a new fraction of grid supply is disconnected (during the interval 20:00-20:15) battery cannot meet demand during that interval and supply failures take place at that time. At this point, the total cost of energy has been reduced to € 66.88 and total emissions have decreased up to 166.0 kg of CO_2. Therefore, to complete iteration i=7, a fraction of grid supply is disconnected during 20:00-20:15 and ESS is charged during the cheapest interval (during 04:00-04:15). The result of this iteration is a total cost of € 65.12 and overall emissions of 164.2 kg of CO_2.

Continuing with this procedure the optimal situation is reached in iteration i=28, when there is no available time interval to disconnect grid supply charging batteries in previous intervals to meet demand and obtain an economic benefit. This situation is shown in Figure 4. In the final situation, the total cost is € 55.97 and the total emissions are 160.8 kg of CO_2.

4. Scenarios Results and Ddiscussion

Table 1 shows the final results of both scenarios. DEROP allows great savings in energy costs in both scenarios. However, DEROP allows lower energy costs in scenario A, where the initial costs are higher. This shows that even in a facility where a tariff with hourly discrimination is not expected to be profitable, DEROP takes advantage of ESSs and allows greater savings. Furthermore, DEROP optimises scenario A faster than B (28 iterations instead of 31). The conclusion is that DEROP works faster with tariffs with hourly discrimination and it achieves better results in this case. This can be observed in Figure 5, which shows the evolution of costs throughout the simulation of both scenarios.

This algorithm has been coded in Matlab and each simulation takes just a few seconds, which supports the quickness of the developed algorithm.

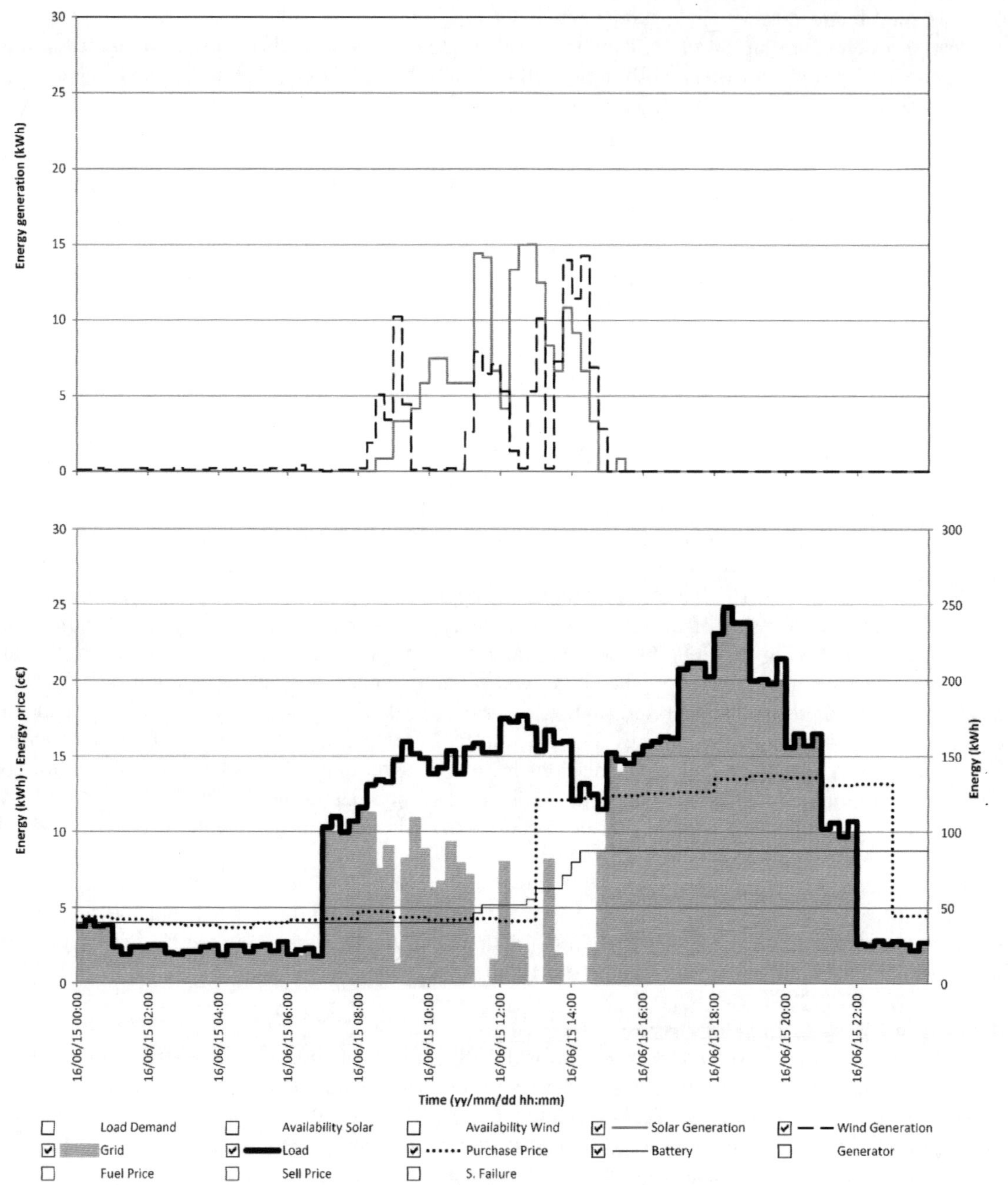

Figure 3. First iteration in scenario A.

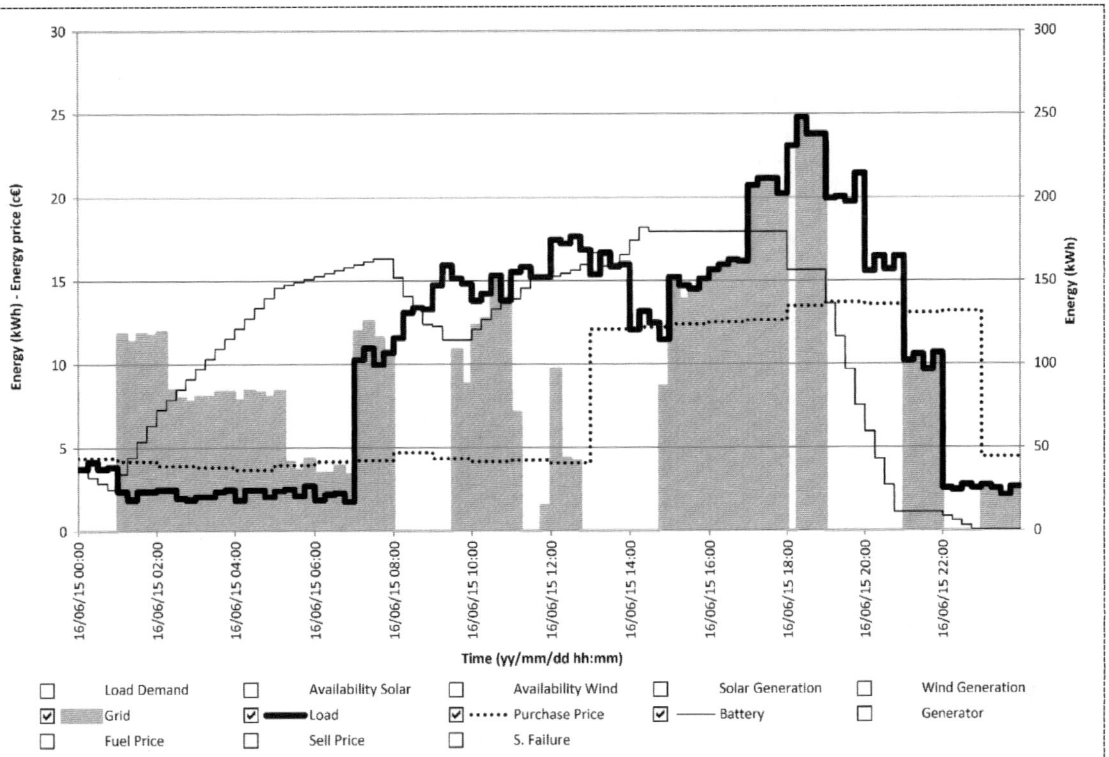

Figure 4. Simulation results in scenario A (tariff with hourly discrimination).

Table 1. Simulation results comparison between both scenarios.

	Scenario A	Scenario B
Cost (€)	55.97	61.14
Initial cost in $S^{(0)}$ (€)	97,14	96,82
Savings compared to $S^{(0)}$ (%)	42.38	36.85
Emissions (kg CO_2)	160.80	162.00
Emissions reduction compared to $S^{(0)}$ (%)	37.57	37.10
Solar generation (kWh)	191.67	191.67
Wind generation (kWh)	133.33	133.33
Demand (kWh)	1033.78	1033.78
Power grid supply (kWh)	669.41	669.57
Grid supply savings compared to $S^{(0)}$ (%)	35.25	35.23
Number of iterations (N)	28	31

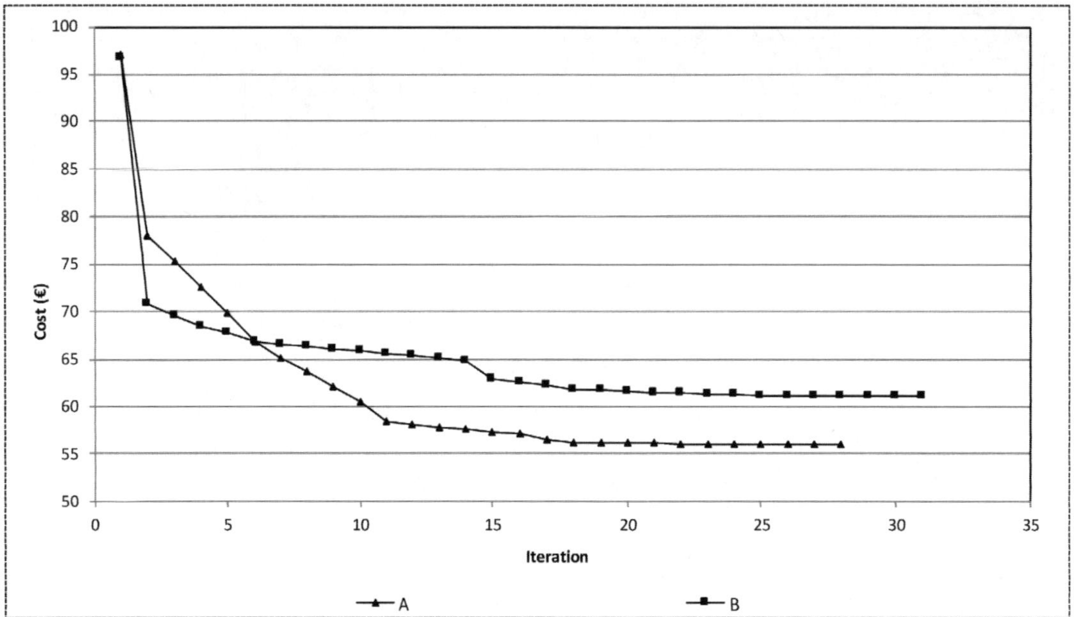

Figure 5. Evolution of costs in both scenarios.

5. Conclusions

In this paper, an optimisation algorithm for DERs management (DEROP) to reach the minimum energy cost in an energy hub is shown.

This method consists of an iterative procedure that uses data related to demand and generation forecasts for each resource, prices of each DER and initial charge state of ESSs to calculate optimal allocation of DERs in an energy hub.

The algorithm has been described and it has been tested in an academic building using real data of prices, demand and generation curves. To test it under different situations, the same building has been optimised with two types of tariff, a tariff with hourly discrimination (scenario A) and a tariff with no time restrictions (scenario B).

The operation of this algorithm in the described scenarios has been shown and the results have been analysed. The simplicity and speed of the presented method must be emphasised. By analysing both scenarios, some interesting results have been observed, such as the great savings that DEROP can reach and the positive impact of these savings both on costs and on emissions. More interestingly, the simulated facility, with low night-time consumption, has higher energy costs with tariffs with hourly discrimination, but thanks to DEROP, an optimal management of DERs allows a final cost 8.46% lower in scenario A than in scenario B. This demonstrates the great capacity of DEROP to optimally manage DERs and to take full advantage of the different tariffs. In addition, it has been demonstrated that DEROP is faster with tariffs with hourly discrimination.

The main advantages of DEROP have been proved with the simulated scenarios. DEROP is simple (its operation has been carefully described in a real example, showing the performance of each iteration), it is fast (each simulation takes just a few seconds when it is coded in Matlab), it is flexible (it allows adding as many resources and constraints as needed, in addition to using non-linear functions for costs and emissions) and it is easy to be implemented in real facilities, as it uses data that can be easily gathered with an energy management system. Moreover, DEROP allows great savings in real facilities through optimising the schedule of DERs and ESSs.

As a final comment, the developed algorithm could be used with other purposes such as minimising CO_2 emissions or optimal schedule of maintenance tasks. In addition, the methodology described in this paper might be used to define the size of RESs and study their amortisation in real facilities.

6. References

[1] Lasseter R H, June 2011, Proceedings of the IEEE, Smart Distribution: Coupled Microgrids, vol. 99, no. 6, pp. 1074–82.

[2] Thomaidis N S, Santos-Alamillos F J, Pozo-Vázquez D and Usaola-García J, February 2016, Computers & Operations Research, Optimal management of wind and solar energy resources, vol. 66, pp. 284–91.

[3] Tianguang L and Qian A, February 2016, Applied Energy, Interactive energy management of networked microgrids-based active distribution system considering large-scale integration of renewable energy resources, vol. 163, pp. 408–22.

[4] Brahman F, Honarmand M and Jadid S, March 2015, Energy and Buildings, Optimal electrical and thermal energy management of a residential energy hub, integrating demand response and energy storage system, vol. 90, pp. 65–75.

[5] Abedini M, Moradi M H and Hosseinian S M, May 2016, Renewable Energy, Optimal management of microgrids including renewable energy sources using GPSO-GM algorithm, vol. 90, pp. 430–9.

[6] Kienzle F, Ahcin P and Andersson G, April 2011, IEEE Transactions on Sustainable Energy, Valuing Investments in Multi-Energy Conversion, Storage, and Demand-Side Management Systems Under Uncertainty, vol. 2, no. 2, pp. 194–202.

[7] Rastegar M and Fotuhi-Firuzabad M, November 2015, Energy and Buildings, Load management in a residential energy hub with renewable distributed energy resources, vol. 107, pp. 234–42.

[8] Morais H, Kádár P, Faria P, Vale Z A and Khodr H M, January 2010, Renewable Energy, Optimal scheduling of a renewable micro-grid in an isolated load area using mixed-integer linear programming, vol. 35, no. 1, pp. 151–6.

[9] Izadbakhsh M, Gandomkar M, Rezvani A and Ahmadi A, March 2015, Short-term resource scheduling of a renewable energy based micro grid, vol. 75, pp. 598–606.

[10] Khalid M, Ahmadi A, Savkin A V and Agelidis V G, November 2016, Minimizing the energy cost for microgrids integrated with renewable energy resources and conventional generation using controlled battery energy storage, vol. 97, pp. 646–55.

[11] Spanish System Operator Website (Red Eléctrica de España): www.ree.es. June 2016.

Acknowledgments

This research work has been made possible with the support of the Programa de Apoyo a la Investigación y Desarrollo (PAID-06-12) de la Universitat Politècnica de València (Spain) and the GV/2015/068-Ayudas para la realización de proyectos de I + D para grupos de investigación emergentes.

Production of Biodiesel Using a Membrane Reactor to Minimize Separation Cost

O A Olagunju[1] and P Musonge[1]

1 Department of Chemical Engineering, Durban University of Technology, South Africa, 4000
E-mail: gilbert4life2004@yahoo.com, paulm@dut.ac.za

Abstract. This study investigates the performance of a packed bed membrane reactor in the transesterification process of triglycerides to methyl ester using soyabean oil as feedstock. A TiO_2/Al_2O_3 ceramic microporous membrane was selected due to its chemical inert nature and thermal stability to selectively remove the product from the reaction medium. CaO impregnated on the surface of activated carbon was packed into the membrane and acted as catalyst. The synthesized catalyst had a total loading of 40.50 % and was characterized by XRD and temperature-programmed desorption of CO_2 (CO_2-TPD). The crude biodiesel produced was micro-filtered by the ceramic membrane with a pore size of 0.02 µm to retain the unreacted oil and free glycerol, at the transmembrane pressure of 100 KPa. The best condition was achieved with a temperature of 65 °C, methanol/oil molar ratio of 6:1 for 150 minutes, which resulted in the highest FAME yield of 94 %. Methyl ester produced met the ASTM D6751 and SANS 1935 specifications. The product obtained was mainly composed of methyl esters. Glycerol was not detected in the product stream due to the ability of the membrane to retain the glycerol and the unreacted oil in the medium, which solved the issue of glycerol separation from biodiesel.

1. Introduction

The continuous exploration of sustainable energy resources has become a global challenge. Biofuel has been identified as one of the most favorable renewable energy sources with great potentials to solve issues relating to global fossil fuel addiction. Biofuels are sustainable, biodegradable and environmentally friendly. Currently, biodiesel is gaining considerable recognition as a sustainable renewable energy source because it is renewable, non-toxic and eco-friendly [1]-[3].

This biofuel has several benefits over conventional fossil fuels, and some of them include cleaner engine emissions, renewability, and better lubricating properties, which makes it a good alternative fuel [4]. Biodiesel can be produced from various sources such as vegetable oil (canola, soybean and sunflower oil), non-edible oils (Jatropha curcas, rapeseed oil) and used vegetable oil [5]-[8].

Biodiesel is mostly produced in the conventional way using homogeneous base catalysts such as sodium hydroxide or potassium hydroxide. Acid homogeneous catalysts have also been used but the rate of reaction is slow compared to the base catalyzed reaction. Homogeneous processes have many shortcomings such as further separation of the catalyst from the reaction mixture, generation of large amounts of wastewater during the separation and purification of the product, and formation of soap due to the reaction of the alcohol with the free fatty acid present in the reaction medium [9]. Baroutian *et al.* [10] stated that biodiesel produced using KOH as the catalyst generated a high yield of the product but an additional process such as purification of product and treatment of wastewater generated was further needed. Previous research conducted by Saleh *et al.* [11] showed that for each litre of biodiesel produced, about 10 litres of wastewater is generated and the treatment invariably increased the cost of production.

Biodiesel may be regarded as a suitable sustainable energy with many advantages but still seems less attractive due to its cost as compared to fossil fuel. In other to make biodiesel more competitive and attractive to users a better approach which eliminates the various challenges associated with its production such as immiscibility of reactants, separation of the product, further purification of product and wastewater generation has to be provided. One of such approaches involves the use of membrane separation technology to explore some of the challenges encountered by using homogeneous transesterification.

Membrane separation technology has been extensively used in wastewater purification by various researchers because of its ability to carry out the separation of different components in a single process stream based on their molecular weight. In this process, reaction and separation take place within a single unit and therefore eliminates the need for a further purification step. Membrane reactors have the ability to selectively remove the product from the reaction mixture by allowing the component with lower molecular weight to pass through its pores while higher molecular weight components are rejected. This method also allows for proper contact between the immiscible reactants and catalyst and as a result, a higher yield of the product can be achieved [12].

South Africa is known to be a water scarce country and the use of membrane technology to produce biodiesel will help to conserve water for other purposes as it will eliminate the use of water for purifying the product and makes the environment free from pollution. Baroutian *et al.* [10] had previously explored this technology, using potassium hydroxide as catalyst and palm oil as feedstock. However, the problem of membrane selectivity remains, which this study aims to address. Furthermore, a comparative study using a different feedstock and catalyst is essential for affirming this technology for the production of biodiesel.

In this study, a packed bed membrane reactor was used to produce FAME. The combination of heterogeneous alkali transesterification and triglyceride separation in the packed bed membrane reactor was employed. The advantage of this new process is in the simplification of the refining steps.

The ceramic membrane is a filter medium composed of inorganic oxides, which permeate liquids through pores and separate out solid matters by retention. Ceramics have high permeability, good separation i.e. sharp cut-off value and long lifetime. Ceramic membranes offer more advantages over the polymeric membranes and these include; high chemical resistance, high thermal stability, and high durability. The chemical stability makes possible, the utilization of chemically aggressive cleaning methods. Furthermore, the thermal stability enables the use of steam and high temperatures during its operation [10].

For this purpose, a tubular ceramic (TiO_2/Al_2O_3) membrane was used as reactor and separator because of its high thermal stability. The tubular membrane was packed with calcium oxide catalyst supported on activated carbon. Among the available supports [13] [14] activated carbon has proved to be highly effective as a catalyst support that can be used in the transesterification process. Activated carbon has a large surface area, inert carbon skeleton, good physical/thermal stability and it can be locally produced at low cost.

Thus, the objective of the present study is to evaluate the applicability of using a packed bed ceramic membrane reactor to separate biodiesel from the reaction medium by analyzing the permeate flux and the quality of the product obtained. The influence of process parameters on the performance of the reactor was also evaluated.

2. Experimental

2.1. Materials

Soybean oil was purchased from a local market. Methanol (99.8%) was supplied by Laboratory supplies co., South Africa. Pure calcium oxide (98.9%) was used as a catalyst for transesterification and was obtained from associated chemical enterprises, South Africa and the activated carbon granules used as the catalyst support in this study was purchased from the same company.

2.2. Reactor Setup

Figure 1 shows the experimental setup for the biodiesel production. A tubular ceramic membrane (Atech Innovations Gmbh, Germany) was used as reactor and separator. The length, inner diameter, outer diameter and pore size of the membrane were 1000 mm, 16 mm, 25.4 mm and 0.02 μm, respectively. The filtration surface area for the entire membrane was 0.0201 m^2. A Watson Marlow 313S peristaltic pump (Cole-Parmer Instrument, USA) was used to feed the reactants into the system. The Chem-Durance chemical resistant pump tubing with a size of 16 (ID = 44 mm, OD = 2.36 mm) was used because of its ability to withstand the operation temperature. A shell and tube heat exchanger equipped with a hot water circulating bath was used as the heat transfer medium. Pressure gauges and thermometers were used to monitor pressures and temperatures of the system. Catalysts particles were packed inside the ceramic membrane and held in place using cloth screens attached to the upstream and downstream tubing. It was observed from the preliminary experiment conducted, that methanol and biodiesel molecules were able to pass through the membrane, which can be attributed to their smaller molecular sizes lower than the pore size of the membrane while the glycerol and unreacted oil go into the recycling tank. This finding quite differs from the result of the study carried out by Baroutian *et al.* [10] who recorded that product, by-product and methanol were able to pass through the membrane, which means the filtrate containing FAME, and glycerol needs a further separation process for pure FAME to be obtained and this can take several hours to be accomplished. Although the pore size of the membrane used for this experiment was 0.02μm and may be the reason for the inability of glycerol to pass through.

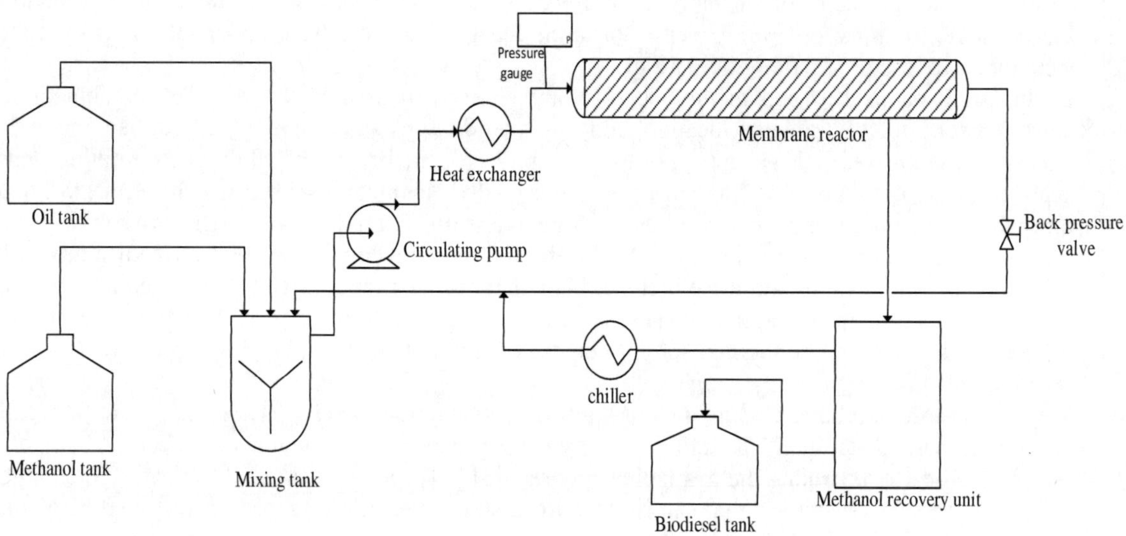

Figure 1. Schematic diagram of a membrane reactor for biodiesel production.

2.3. Catalyst Preparation and Characterization

The catalyst solution was prepared by dissolving the CaO in deionized water. Activated carbon was sieved to size range from 450 to 810 μm, washed with deionized water to remove fines and dirt, oven dried at 100 ^{0}C for a day, cooled in a desiccator and stored in glass bottles. Activated carbon was added to the catalyst solution and then agitated in an orbital shaker at 180 rpm at a constant temperature of 25 ^{0}C for 24 h. The amount of adsorbed CaO was measured by the gravimetrical method. The total loading content of CaO was 40.50 % by weight, based on the initial weight of activated carbon.

In addition, the characteristics of the prepared supported catalyst were determined. Scanning electron micrographs (SEM) were obtained on an FEI Quanta 200 FESEM scanning electron microscope. The accelerating voltage was 20 kV. The SE and BSE detectors were ETD and Low kV SSBSED, respectively. To evaluate the specific surface area and pore volumes, adsorption of nitrogen at 77 K was carried out by ASAP 2020, Micromeritics. Prior to taking adsorption data, degassing at 120 ^{0}C and a residual pressure of 300 μmHg for 24 hours was performed using the degassing port.

2.4. Transesterification in the Packed Bed Membrane Reactor
Soybean oil and methanol were charged into the mixing tank separately. The volume ratio of oil to methanol was varied between 3:1 to 12:1 and the catalyst was packed into the membrane reactor. Methanol was charged continuously into the reactor using the circulating pump according to Olagunju and Musonge [15] and heat exchanger was started up to heat the reactants. Subsequently, the reactor was filled with the reactant. The pressure inside the membrane was monitored by two pressure gauges and was controlled at 100Kpa. The permeate stream containing biodiesel and methanol was collected in the beaker. After each run, the circulating pump and heat exchanger were switched off. Thereafter, the system was fully drained, the catalysts were taken out and the system was flushed for 30 min with methanol and then drained. Biodiesel yield in the present experiment was calculated by using the following equation (1):

Biodiesel conversion (%) = (Mass of the biodiesel / Mass of the oil used) * 100% (1)

3. Results and Discussion

3.1. Catalyst Properties
The SEM analysis of calcium oxide catalyst supported on activated carbon (CaO/AC) showed a good dispersion of calcium oxide on the surface of activated carbon as shown in Figure 2. Based on this result, after loading the catalyst, activated carbon retained its structure and the calcium species were found vastly distributed upon the surface of the support.

Table 1. Characterisation of CaO/AC catalyst

Characteristic	Value	Units	Technique
BET surface area	240.51	m^2/g	BET
Pore volume	0.152	cm^3/g	BET
Micro pore volume	0.121	cm^3/g	BET
Average pore width	2.87	nm	BET
Active sites concentration	1.436	mmol/g	TPD-CO_2

Surface area, pore volume and pore width of the supported catalyst on activated carbon are presented in Table 1. The significant reduction in BET surface area from virgin activated carbon (1425 m^2 /g) to the CaO/AC catalyst with 40.50 wt. % loading (240.51 m^2 /g) indicates filling of calcium oxide molecules into the activated carbon pores. CO_2 temperature programmed desorption (TPD) method was used to determine the basicity of the catalyst.

3.2. Selection of Suitable Membrane for Separation
In order to achieve the suitability of membrane selection, it is important to estimate the size of the dispersed oil droplets in the continuous alcohol phase. The minimum particle size in the oil-methanol emulsion can be estimated from the work of DeRoussel *et al.* [16]. According to previous studies, the average drop size for the unreacted oil was 44 microns with a lower and upper size limit of 12 and 400 microns, respectively [16]. The membrane pore size selected in this study is 0.02 microns, which was able to trap the unreacted oil within the membrane.
The retention of free glycerol and unreacted oil in the reaction medium micro-filtrated by the 0.02 μm membrane was over 99%, which met the standard for biodiesel. The content of impurities was 2 mg/kg and was better than the results of water washing process, so this membrane pore size was selected as a suitable one for the biodiesel refining.

Figure 2. SEM image of calcium oxide supported on activated carbon (CaO/AC).

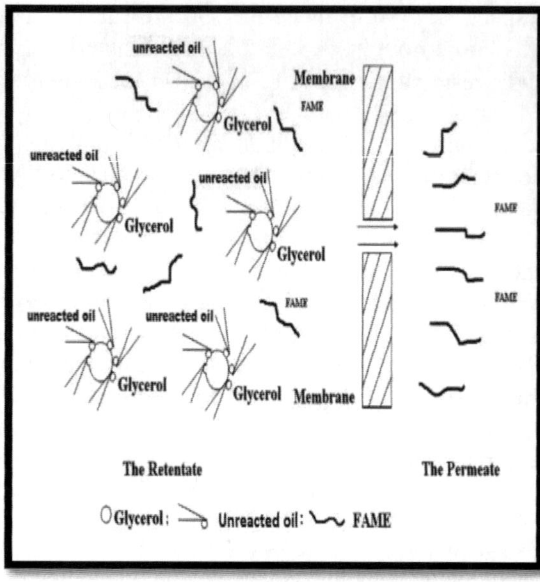

Figure 3. Cross-filtration of crude biodiesel by membrane separation.

The presence of minor residual soap and free glycerol in the biodiesel can cause a serious engine problem and hazardous emissions [17]. Although the South African biodiesel standard (SANS 1935) does not set a direct property for the soap in the final product, the limitation of the content of contaminants in the biodiesel makes a strict rule for the soap content in the refined biodiesel so as to meet the limitation of free glycerol content of less than 0.02 (wt %)

The principle of membrane separation of unreacted oil and free glycerol is depicted in Figure 3. Due to the immiscibility of free glycerol and biodiesel and surface activity of soap, the unreacted oil existed in the form of reversed micelle, which was very similar to the form of phospholipids in the hexane micelle whose size was larger than a single solute molecule [18]. The free glycerol existed in form of droplets suspended in the reaction medium. The hydrophilic end of unreacted oil bond to the free glycerol droplets, while the hydrophobic ones immersed into the crude biodiesel.

3.3. Separation of the Permeate Stream

When the experiments were carried out, the permeate from the membrane reactor percolated as a clear, amber, homogeneous solution. As the solution began to reach room temperature, the permeate separated into two phases with a clear boundary. The absence of particles in the permeate enhanced the separation process. GC analysis revealed that the homogeneous samples were those with the lowest percentage of FAME.

Table 2. Composition of the lower and upper layer for permeate samples taken at 30 and 90 minutes from the beginning of permeation

Phase	FAME (wt. %)	Glycerol (wt. %)	Methanol (wt. %)	Calculated Density (g/mL)
Lower (30 min)	85.5	0	14.5	0.863
Upper (30 min)	19.5	0	80.5	0.828
Lower (90 min)	93.4	0	6.6	0.861
Upper (90 min)	15.6	0	84.4	0.825

Table 2 shows the composition of the different phases. The lower phase in both of these samples was the FAME-rich phase, with an average mass composition of 85.5% after 30 minutes and 93.4% after 90 minutes from the start of permeation (run with membrane pore size 0.02 micron and 300 g oil injection). All samples of the FAME-rich phase taken throughout the run had non-detectable levels of glycerol using GC. In the results, the upper phase was found to be the methanol-rich phase containing no glycerol. This indicates that the amount of FAME in the FAME-rich phase was such that its density was greater than the density of the methanol-rich phase. One important finding was that the location of the FAME-rich phase and methanol-rich phase changed due to no glycerol content. The recycling of the methanol-rich phase is of interest, in order to maximize its usage.
The above observations indicate that the methanol-rich phase of the de-phased permeates can be recycled to the reactor in order to decrease the overall methanol: oil molar ratio. In order to reach a commercial methanol to oil molar ratio of 6:1, it is entirely practical to cool the permeate to allow for phase separation and recycle the methanol-rich phase to the reactor. According to a previous study carried out by Olagunju and Musonge [15], methanol: oil mole ratios of 3:1, 6:1, 9:1 and 12:1 were used and the best condition was achieved with a temperature of 65 °C, methanol/oil molar ratio of 6:1 for 150 minutes, which resulted in the highest FAME yield of 94 %. These operating parameters and yield served as basis for this current experiment study.
At the completion of the runs, the membrane reactor retentate was observed to contain a stable interfacial emulsion similar to that obtained in a batch reactor run under identical process conditions. The conventional method requires several hours to achieve complete separation of the FAME-rich and methanol/glycerine-rich phases. This is because emulsions from the oil or soaps produced during the reaction process may act as emulsifiers or dispersants to prevent phase separation. The situation in a batch reactor would even be worse when using high free fatty acid feedstock. In that case, a significant reduction in biodiesel yield would result. The membrane is thus performing an additional purification function by retaining emulsions within the reactor. The membrane reactor was also tested as a batch reactor without pressurizing the system. The entire content in the reactor was then transferred into a separating funnel and effective separation was completed after 5 hours.

3.4. FAME Characterization
Physical and chemical properties of the produced biodiesel were measured according to the test methods recommended by the American Society for testing and Materials (ASTM). The results of these characterizations are listed in Table 3.

Table 3. Characteristics of FAME produced from soya bean oil using membrane reactor

Characteristic	Test	Units	SANS 1935 Specification Limit	Result
Density @ 15°C	ASTM D7042	g/mL	0.86-0.9	0.87
Viscosity @ 40 °C	ASTM D7042	cSt	3.5-5	3.8
Flash point	ASTM D93	°C	120 min	167
Water content	ASTM D6304	%	0.05 max	-
Total acid number	-	mgKOH/g	0.5 max	0.21
Total Contamination	IP 440	mg/Kg	24 max	2
Sulphur	ASTM D4294	ppm	10 max	1

The results show that the FAME produced using membrane technology is within ASTM standard specifications and biodiesel specification. Biodiesel properties are numerous but the most important properties are those that have direct impact on the performance of the engine such as viscosity, flash point, density and so on. All these properties help to increase the lifespan of the engine, give a better lubrication and complete combustion so that the engine can produce a higher energy output.

4. Conclusion

The experimental setup for this project was constructed and the membrane reactor exhibited a good performance in the transesterification of the high quality of biodiesel yield, which does not require any additional purification process. The following conclusions could be drawn from the study:

(1) The ceramic membrane with the pore size of 0.02 μm was very suitable for this reaction and separation process due to its high flux and the good quality of the permeate.

(2) This new method of biodiesel production showed the advantage of no water usage in the process as compared to the conventional water washing that leads to wastewater being generated and consequently leading to environmental pollution requiring additional treatment.

(3) The characteristics of the product under these conditions were within the ASTM standard.

5. Reference

[1] Jiang J J and Tan C S 2012 Biodiesel production from coconut oil in supercritical methanol in the presence of co-solvent *J. Taiwan Inst. Chem. Eng.* 43 102–7

[2] Manh D V, Chen Y H, Chang C C, Chang M C and Chang C Y 2011 Biodiesel production from tung oil and blended oil via ultrasonic transesterification process *J.Taiwan Inst. Chem. Eng.* 42 640–44

[3] Zhou C, Wang C, Wang W, Wu Y, Yu F, Chi R and Zhang J 2010 Continuous production of biodiesel from soybean oil using supercritical methanol in a vertical tubular reactor: I. Phase holdup and distribution of intermediate product along the axial direction *Chin. J. Chem. Eng.* 18 626–9

[4] Gan M, Pan D, Ma L, Yue E and Hong J 2009 The kinetics of the esterification of free fatty acids in waste cooking oil using $Fe_2 (SO_4)_3/C$ catalyst *Chin. J. Chem. Eng.* 17 83–7

[5] Alcantara A, Amores J, Canoira L, Fidalgo E, Franco M J and Navarro A 2000 Catalytic production of biodiesel from soybean oil used frying oil and tallow. *Biomass and Bioenergy* 18 515–27

[6] Leung D and Guo Y 2006 Transesterification of neat and used frying oil: optimization for biodiesel production *Fuel Processing Technol.* 87 883–90

[7] Mittelbach M and Remschmidt C 2004 *Biodiesel-The comprehensive handbook*, First ed., Boersedruck, Ges.m.b.H, (Austria)

[8] Wang Z M, Lee J S, Park J Y, Wu C Z and Yuan Z H 2007 Novel biodiesel production technology from soybean soap stock. *Korean J. Chem. Eng.* 24 1027–30

[9] Sharma Y C and Singh B 2010 Development of biodiesel: current scenario *Renewable and Sust. Energy Rev.* 13 1646-51

[10] Baroutian S, Aroua M K, Raman A A A and Sulaiman N M N 2011 A packed bed membrane reactor for production of biodiesel using activated carbon supported catalyst *Bioresource technol.* 102 1095–1102

[11] Saleh J, Tremblay AY and Dubé M A 2010 Glycerol removal from biodiesel using membrane separation technology *Fuel* 89 2260–6

[12] Westermann T and Melin T 2009 Flow-through catalytic membrane reactors principles and applications *Chem. Eng. Process.* 48 17–28

[13] Xie W and Huang X 2006 Synthesis of biodiesel from soybean oil using heterogeneous KF/ZnO catalyst *Catal. Lett.* 107 53–59

[14] Zabeti M, WanDaud W M A and Aroua M K 2009 Optimization of the activity of CaO/Al_2O_3 catalyst for biodiesel production using response surface methodology *Appl.Catal. A* 366 154-9

[15] Olagunju O A and Musonge P 2015 The performance of a packed bed membrane reactor for the production of biodiesel 2nd Intl., Conf., on Composites, Biocomposites & Nanocomposites 28-30 October 2015 Durban University of Technology, South Africa

[16] DeRoussel P, Khakhar D V and Ottino J M 2001 Mixing of viscous immiscible liquids. Part 2: over emulsification interpretation and use *Chem. Eng.* 56 5531

[17] Plank C and Lorbeer E 1995 Simultaneous determination of glycerol and mono- di- and triglycerides in vegetable oil methyl esters by capillary gas chromatography *J. Chromatogra. A* 697 461–8

IOP Conf. Series: Earth and Environmental Science **78** (2017) 012019 doi:10.1088/1755-1315/78/1/012019

[18] Lin L, Rhee K C and Koseoglu S S 1997 Bench-scale membrane degumming of crude vegetable oil: process optimization *J. Membr. Sci.* 134 101–8

Acknowledgements
The financial support from the Institute of Wastewater treatment, Durban University of Technology, South Africa is highly acknowledged.